# Wireless Communications and Mobile Commerce

Nan Si Shi
University of South Australia, Australia

**IDEA GROUP PUBLISHING**
Hershey • London • Melbourne • Singapore

| | |
|---|---|
| Acquisitions Editor: | Mehdi Khosrow-Pour |
| Senior Managing Editor: | Jan Travers |
| Managing Editor: | Amanda Appicello |
| Development Editor: | Michele Rossi |
| Copy Editor: | Lori Eby |
| Typesetter: | Jennifer Wetzel |
| Cover Design: | Michelle Waters |
| Printed at: | Integrated Book Technology |

Published in the United States of America by
  Idea Group Publishing (an imprint of Idea Group Inc.)
  701 E. Chocolate Avenue, Suite 200
  Hershey PA 17033
  Tel: 717-533-8845
  Fax: 717-533-8661
  E-mail: cust@idea-group.com
  Web site: http://www.idea-group.com

and in the United Kingdom by
  Idea Group Publishing (an imprint of Idea Group Inc.)
  3 Henrietta Street
  Covent Garden
  London WC2E 8LU
  Tel: 44 20 7240 0856
  Fax: 44 20 7379 3313
  Web site: http://www.eurospan.co.uk

Library of Congress Cataloging-in-Publication Data

Wireless communications and mobile commerce / Nansi Shi, editor.
    p. cm.
Includes bibliographical references and index.
  ISBN 1-59140-184-4 -- ISBN 1-59140-185-2 (ebook)
  1. Mobile commerce. 2. Wireless communication systems. I. Shi, Nan
Si, 1953-
  HF5548.34.W57 2003
  658.8'72--dc22
                                    2003014945


British Cataloguing in Publication Data
A Cataloguing in Publication record for this book is available from the British Library.

All work contributed to this book is new, previously-unpublished material. The views expressed in this book are those of the authors, but not necessarily of the publisher.

# Wireless Communications and Mobile Commerce

## Table of Contents

# Preface

The emerging wireless communications and mobile computing have been merging with the Internet, and this phenomenon has made possible the introduction of electronic commerce to a new business application and research subject: mobile electronic commerce or mobile commerce (m-commerce).

Although m-commerce can be generally considered to be an extension of electronic commerce, it has a number of unique characteristics and inherent complexities, as it embraces many emerging technologies. This book seeks to address and explore these unique issues associated with the phenomenon, in order to provide a better understanding regarding the current situation, its growing trend, practice, and further studying guidelines in this emerging area.

It is not feasible for a single author to write a book of this type and cover all of the important aspects of this rapidly emerging area, within a required time frame, and yet maintain a degree of depth and width. One of the better approaches is to invite related experts around the world to contribute their knowledge.

Therefore, we identified some of the key topics in this area and invited a wide range of professionals across the globe to contribute a chapter in their areas of expertise. Experts in various related disciplines from different parts of the world responded to this project enthusiastically. We gave adequate time for completing the project, spending months preparing and consulting with the publishers, authors, and reviewers, in order to avoid a hasty approach and to yield qualified outcomes.

The success of this book is largely due to the collective efforts of the wonderful team consisting of the authors and other reviewers. The inclusion of the authors in the blind review process improved the quality of the book and also served as an incentive to each author to strengthen his or her write-up. Although the editor initially received many proposals and manuscripts, the stringent quality control measures taken permitted us ultimately to include only 10 chapters, contributed by experts in different parts of the world: Canada, Finland, Portugal, Singapore, South Africa, the United Kingdom, and the United States. The team not only includes experts from Information and Communications Technology (ICT) areas, like Faculty of Computer Science, Information Systems, Telecommunication Networks, and Software Engineering and Information Technology, but also includes professionals from Business and Management fields, such as colleges of business administration, faculties of management, graduate schools of management, and other kinds of business schools. Some chapters are outcomes of joined forces with both ICT and business experts, presenting cross perspectives and knowledge.

# READERSHIP

The primary readers of this book are professionals, researchers, executives, graduate and postgraduate students in ICT and business areas, as well as others with strong interest in wireless communications and m-commerce. Readers will find that this book is a rich, informative introduction for obtaining an overall understanding of wireless communications and m-commerce.

This is mainly because this book consists of qualified study outcomes from multiple disciplines. These rich research findings, holistic perspectives, suggestions, descriptions, conclusions, recommendations, arguments, and references presented by these authors will assist readers in pursuing further studies in many particular aspects.

# ORGANIZATION

This book is organized into three sections, with the following major themes:

1. Global Mobile Commerce—Unique Characteristics, Opportunities, Challenges and Limitations
2. Mobile Business Network and Consumers
3. Wireless Communications and Mobile Computing Infrastructure Considerations

# OVERVIEW

As the book is devoted to a very diverse range of topics written by a large number of professionals and academics, it is necessary for the editor to provide a bird's eye view of the contents of the chapters.

*Section I* deals with *Global Mobile Commerce—Unique Characteristics, Opportunities, Challenges and Limitations* and it consists of four chapters.

Chapter 1, *Global Heterogeneity in the Emerging M-Commerce Landscape,* by Nikhilesh Dholakia, Ruby Roy Dholakia, Mark Lehrer, and Nir Kshetri, provides a framework, derived from macrodata and selected case studies, for understanding cross-national and cross-regional variations in the evolution of m-commerce applications. This chapter also (a) provides background on the rapid diffusion of mobile technology; (b) examines the global diffusion pattern of mobile phones and m-commerce technology; (c) identifies a variety of economic, social, and technological factors impacting the diffusion patterns of mobile technology and m-commerce; and finally, (d) develops a simple typology for tracking future developments in m-commerce.

In Chapter 2, *Assessing the Market Potential of Network-Enabled 3G M-Business Services,* Mats Samuelsson and Nikhilesh Dholakia argue that the key differentiators of m-business are made up of a set of the Experience and Function parameters that set mobile offerings (m-offerings) apart from e-business. The authors not only present a constellation of m-business service offerings that could take advantage of new network-enabled services, but they also suggest an approach to add value and differentiate m-service offerings so that businesses can continue to remain profitable. Using an evaluative framework that they developed, the authors offer a comparative assessment of the profiled m-business services and general guidelines for assessing new m-business services.

Chapter 3 by Mahesh S. Raisinghani on *Mobile E-Commerce and the Wireless Worldwide Web: Strategic Perspectives on the Internet's Emerging Model,* provides an overall description of m-commerce and examines its current, state-of-the-art opportunities and challenges and future trends. The authors of this chapter argue that there are many new opportunities that have only begun to be explored, but like other capital ventures, these new opportunities have their drawbacks that may limit growth of the m-commerce market if not dealt with. The requirements for operating m-commerce and the numerous ways to provide wireless business are explained.

In Chapter 4, *Opportunities and Limitations in M-Commerce,* P.W. Lei, C.R. Chatwin, R.C.D. Young, and S.H. Tóng examine the opportunities and limitations in m-commerce and concentrate their discussion on mobile phone systems. The authors argue that m-commerce has its own unique characteristics and functionality, so it creates unique and new business opportunities, but with limitations.

*Section II* discusses issues regarding *Mobile Business Network and Consumers* and comprises three chapters.

In Chapter 5, *Understanding Emergent M-Commerce Services by Using Business Network Analysis: The Case of Finland*, Tommi Pelkonen and Nikhilesh Dholakia discuss complex networks of business relationships, comprising telecommunications service providers, mobile device makers, financial linkage providers, and various third-party value-adding companies. Developed in this chapter is a framework from illustrative analyses of the Finnish situation, and then general guidance is given for the formation and sustenance of effective business networks for m-commerce players worldwide.

Provided in Chapter 6, *Understanding the Mobile Consumer*, by Constantinos Coursaris, Khaled Hassanein, and Milena Head, is an analysis of the emerging global m-commerce market from a consumer's perspective. Presented in this chapter is a consumer-centric m-commerce model outlining the various wireless interaction modes of the m-consumer and discussing the needs and concerns of the m-consumer. Also presented is an m-commerce value network outlining the roles of the different players within this industry and various business applications developed to address the needs of the m-consumer. Finally, a global m-commerce market overview is provided, and some future trends are outlined.

In Chapter 7, *Intelligent Product Brokering and User Preference Tracking*, by Sheng-Uei Guan, Chon Seng Ngoo, and Fangming Zhu, a design for an evolutionary ontology-based product-brokering agent is proposed, one potential application in the area of m-commerce. In most current systems, user-supplied keywords are normally used to generate a profile for the user. The proposed solution uses an evaluation function to represent the user's preference instead of the usual keyword-based profile. By using genetic algorithms, the agent tries to track the user's preferences for a particular product by tuning some parameters inside. The authors implemented a prototype in Java, and the results obtained from their experiments look promising.

*Section III* is concerned with *Wireless Communications and Mobile Computing Infrastructure Considerations* and is composed of three chapters.

Presented in Chapter 8, *Directions in Wireless Telecommunications: Analytical and Operational Pathfinders*, contributed by John H. Nugent, are high-level analytical and operational tools and models that assist the wireless telecommunications professional in understanding the telecommunications market's characteristics, life cycles, trends, directions, limits, and drivers. Tools are also presented that provide important and timely insight for gaining competitive advantage based upon early detection of critical inflection points.

In Chapter 9, *Wireless Middleware*, by Ken MacGregor, Nico de Wet, Bonnie Lam, and Nadim Yazdani, wireless middleware is introduced as a means of writing distributed applications for mobile environments. Introduced are the concepts of middleware and the additional challenges that arise from wireless communications, in particular, low bandwidth and unreliability. Then described are the commercial wireless products currently available, with particular emphasis on the manner in which they solve the challenges. Finally, introduced in this chapter is the basis for most wireless middleware products—the Java Messaging System (JMS)—and by means of an example, shown is how the JMS can be used to implement a wireless application.

In Chapter 10, *Usability Issues and Limitations of Mobile Devices*, by Suliman Al-Hawamdeh, usability issues and limitations of mobile devices, like limited memory, limited processing power, different technologies and standards, small keyboards, and small screens, are reviewed. Based on findings of a usability study carried out in Singapore, the author suggests that while mobile devices are becoming increasingly popular with the younger generation, users still prefer to use desktops for e-commerce transactions, mainly because of the limitations of mobile devices and the stability and security of wireless networks.

# Acknowledgments

The accomplishment of this book relied on many people's contributions and assistance. It is my pleasure to acknowledge with gratitude the insights and excellent contributions provided by all of the authors. Thanks are also due to other reviewers, including Yufei Yuan from McMaster University (Hamilton, Ontario, Canada) and Mats Arvedson and Ola Eriksson from the Royal Institute of Technology (Sweden).

Special thanks also goes to the staff at Idea Group Publishing, particularly to Mehdi Khosrow-Pour, Jan Travers, Michele Rossi, Amanda Appicello, Carrie Skovrinskie, and Jennifer Sundstrom.

I would like to acknowledge all of the people who encouraged me in this project, especially, Professors Kevin O'Brien and Rod Oxenberry, and Associate Professor Graham Arnold from University of South Australia, Dr. James Brancheau, Dr. David Bennett, Ms. Marilyn Ling, and Mr. Andrew Chen.

Finally, I would like to express my deepest gratitude to my family members for their love and support throughout this project.

*Nan Si Shi, Ph.D., Editor*
*April 2003*

# SECTION I:

# GLOBAL MOBILE COMMERCE—
# UNIQUE CHARACTERISTICS,
# OPPORTUNITIES, CHALLENGES
# AND LIMITATIONS

**Chapter I**

# Global Heterogeneity in the Emerging M-Commerce Landscape

Nikhilesh Dholakia, University of Rhode Island, USA

Ruby Roy Dholakia, University of Rhode Island, USA

Mark Lehrer, University of Rhode Island, USA

Nir Kshetri, University of North Carolina at Greensboro, USA

## ABSTRACT

*Mobile phones, mobile Internet access, and mobile commerce (m-commerce) are growing much faster than their fixed counterparts. Several characteristics of mobile networks make them more attractive than fixed networks for less-developed countries and for those countries that want to "leapfrog" the leading IT nations. To exploit the new mobile communications infrastructures, companies from developed as well as developing countries are rapidly integrating m-commerce technology into their business models. Countries around the world, however, exhibit considerable heterogeneity in their adoption of mobile phones and m-*

*commerce technology. Examined in this chapter is the current stage of mobile technology and m-commerce diffusion across the world, and analyzed are factors influencing the diffusion process. In this chapter, the ways in which the m-commerce landscape of a nation—defined by the penetration rate of mobile phones, the specific combinations of different generations of mobile technology, and the blending of various standards within a given generation—is shaped by politicoeconomic, sociocultural, and policy-related factors are reviewed.*

# INTRODUCTION

The rapid global diffusion of mobile telephones in the last two decades of the 20th century laid the foundation for a new type of technology-aided commerce. Going beyond the computer-mediated electronic commerce (e-commerce)[1] of the 1990s, this new type of mobile commerce (m-commerce) was characterized by novel, location-based services delivered by a variety of handheld terminals (Dholakia & Dholakia, 2003). By 2000, Japan's NTT DoCoMo had already established a huge network of m-commerce service providers and users that relied on that company's i-mode® platform (Bradley & Sandoval, 2002).

Just as networked computers and browser-accessible content provided the preconditions for the takeoff of e-commerce, mobile telephones that are data-ready and connected to digital communications networks provide the preconditions for m-commerce. In recent years, other handheld mobile devices, such as Personal Digital Assistants (PDAs) and enhanced alphanumeric communicators (such as Blackberry devices), supplemented mobile telephones. An increasingly diverse array of such mobile devices is anticipated from the research laboratories of telecommunications, computer, and electronic firms.

Drawing on comparative macrodata from about 30 selected countries, in this chapter, the heterogeneity of factors that influence cross-national differences in the adoption of mobile communication infrastructure and m-commerce applications is explored. It will be seen, in particular, that the distinction between developed, newly developed, and developing countries does not translate into a corresponding continuum of national leaders and laggards in mobile technology adoption. Instead, comparative analysis reveals interesting features of the emergent global m-commerce landscape:

- Not all global leaders in land-based telecommunications or Internet access are the global leaders in mobile connectivity.
- In terms of m-commerce applications, multiple dominant designs are likely to coexist and compete for an extended period of time. This is different from most IT fields, where dominant designs converged rapidly to one or two standards.
- Sources and reasons for national leadership in the evolution of m-commerce applications are likely to be significantly different from the national leadership patterns in Internet, land-line telecommunications, and computers.

Provided in this chapter is a framework, derived from macrodata and selected case studies, for understanding cross-national and cross-regional variations in the evolution of m-commerce applications. In the remainder of the chapter: (a) background on the rapid diffusion of mobile technology will be provided; (b) the global diffusion pattern of mobile phones and m-commerce technology will be examined; (c) a variety of economic, social, and technological factors impacting the diffusion patterns of mobile technology and m-commerce will be identified; and (d) a simple typology for tracking future developments in m-commerce will be developed.

# BACKGROUND: RAPID DIFFUSION OF MOBILE TECHNOLOGY

Mobile technologies—for communications, for accessing the Internet, and for m-commerce transactions—are growing hyperbolically. Mobile devices have become the fastest adopted consumer products to date. In 2000, more mobile phones were shipped than automobiles and PCs combined (Chen, 2000; de Haan, 2000). In the 1990s, the number of mobile phones worldwide grew by 50% annually compared to less than 10% growth in fixed connections. The proportion of mobile phones increased from one out of 50 phone connections in 1990 to one out of three in 1999 (Wellenius, Braga, & Qiang, 2000). The number of mobile subscribers worldwide increased from 11 million in 1990 to 318 million in 1998 (Wai, 2001) and is projected to reach 1.2 billion by 2005, with 450 million using some sort of location-based service (Secker, 2001). The fixed telecommunications industry took more than 130 years to reach comparable levels of diffusion. An estimate suggests that by 2009, there will be more cellular subscribers in the world than fixed-line subscribers (ITU, 2000). Another estimate suggests that by 2005, Internet access through wireless devices will outstrip access via personal computers (UNDP, 2001). Furthermore, more than 25% of e-commerce will take place over handheld sets by 2005 (Shaffer, 2000).

To exploit the opportunity created by the meteoric growth of mobile phones, companies around the world are rapidly integrating m-commerce technology into their business models. This is happening as much in developed as in developing countries, often with fascinating cross-border differences in the most prominent initial applications. An exemplary developed-country m-commerce provider is the U.S. online book retailer Amazon.com that signed deals with wireless providers Sprint PCS, Verizon, Airtouch, and Nextel to leverage m-commerce technology in the company's offerings (Lindsay, 2000). Of fast-moving m-commerce firms in developing countries, a notable example is GWCom. This mobile wireless applications services provider in China launched its wireless portal byair.com in 1998 to provide timely information and e-commerce capabilities, such as stock trading and banking, to users with mobile phones or wireless palmtop devices. By March 2000, byair.com had more than 6,000 subscribers with up to 3,500 daily stock trades and 250,000 page views. It handled more than 30 million information requests and 200,000 wireless stock transactions[2] (Ebusinessforum.com, 2000). Wireless users have been using GWCom's application platforms to conduct online trading since 1998 in Shanghai and since 1999 in Shenzhen. In March 2000, 3,000 investors in Shanghai and 100 in Shenzhen were trading stocks over the paging networks managed by GWCom. The average daily volume of 3,000 Shanghai users in early 2000 was $3.6 million, about 30 times as much as the average trading volume on stockstar.com, the largest and most popular Web-based stock-trading company.

Farmers and small business owners from developing countries are also making extensive use of m-commerce applications. They are utilizing the information gathered via mobile phones to eliminate or reduce the role of intermediaries in the value chain and to lower the risk of their profit margins being squeezed by larger firms or firms from developed countries. Bangladeshi farmers employ such phones to find the proper prices of rice and vegetables. Farmers in remote areas of the Ivory Coast share mobile telephones so that they can follow hourly fluctuations in coffee and cocoa prices in the international market. Thanks to mobile phones, they can now choose the time to sell their crops, when the world prices are in their favor. A few years ago, the only way to find out about market trends was to go to the capital city. Then, the deal making was largely based on often-unreliable information from buyers (Lopez, 2000). Similarly, fishermen in India use mobile phones to obtain information about the price of fish at various accessible ports before deciding where to land their catch (Rai, 2001).[3]

# PATTERNS OF GLOBAL DIFFUSION OF M-COMMERCE TECHNOLOGY

While the statistics above indicate hyperbolic growth and proliferating applications of mobile technologies, global diffusion patterns vary widely. First, uses of mobile phones vary between low- and high-income countries. People in the developing countries often use mobile phones because these may be the only kind of phones available readily (Wooldridge, 1999); and in many regions of Eastern Europe, the mobile phone network is often much more technologically advanced than the older fixed-line network. Thus, while mobile phones represent *supplements* to fixed telephones in high-income economies, they are often *substitutes* for fixed telephones in lower-income economies (ITU, 1997). Likewise, whereas 3G mobile phones provide mobility and efficiency to the users from advanced countries, they are likely to give a large proportion of people from the developing world their first access to the Internet (Banks, 2001). In other words, the developing countries are gravitating to mobile phones because of infrastructure issues (Zuckerman, 2000). In some developing countries, such as Cambodia and Venezuela, mobile penetration already exceeds fixed-line penetration.

For this and for additional reasons examined below, global leaders in mobile technology and m-commerce are not necessarily the richest economies or the leaders in fixed telecommunications and the Internet. Compared in Table 1 are 25 major countries in terms of mobile penetration, fixed-line penetration, Internet usage, and per capita income. Portugal and Taiwan, for instance, have incomes that are less than one third of the incomes of Japan, Switzerland, and the United States, but they are far ahead in terms of mobile penetration. Similarly, Italy has one of the lowest rates of fixed-line penetration (Table 1) but ranks fourth in terms of mobile penetration. Hong Kong, the economy with the highest mobile penetration in the world, has a relatively low rank in terms of Internet usage. As indicated in Table 1, whereas income, fixed telephone penetration, and Internet penetration are significantly correlated with each

## Table 1: Mobile and Fixed Penetrations, Per Capita GNP, and Internet Use

| Country | 2000 Per Capita US$ | | 1999 Fixed Phones | | 1999 Internet Users | | 1999 Mobile Phones | |
|---|---|---|---|---|---|---|---|---|
| | Per Capita GNP | Rank Order | Per 1,000 | Rank Order | Per 1,000 | Rank Order | Per 1,000 | Rank Order |
| Hong Kong | 25,950 | 6 | 576 | 10 | 205 | 14 | 726 | 1 |
| Finland | 24,900 | 10 | 552 | 13 | 404 | 2 | 667 | 2 |
| Sweden | 26,780 | 5 | 665 | 4 | 445 | 1 | 578 | 3 |
| Italy | 20,010 | 18 | 462 | 21 | 158 | 18 | 528 | 4 |
| Taiwan | 16,100 | 19 | 588 | 7 | 216 | 11 | 521 | 5 |
| Austria | 25,220 | 7 | 472 | 20 | 203 | 15 | 519 | 6 |
| South Korea | 8490 | 24 | 438 | 22 | 213 | 12 | 504 | 7 |
| Denmark | 32,020 | 4 | 685 | 2 | 394 | 3 | 499 | 8 |
| Singapore | 24,740 | 11 | 482 | 18 | 289 | 6 | 475 | 9 |
| Portugal | 11,060 | 23 | 424 | 23 | 80 | 24 | 468 | 10 |
| Japan | 34,210 | 3 | 558 | 12 | 162 | 17 | 449 | 11 |
| The Netherlands | 25,140 | 8 | 606 | 6 | 258 | 8 | 435 | 12 |
| Switzerland | 38,120 | 1 | 699 | 1 | 234 | 10 | 420 | 13 |
| United Kingdom | 24,500 | 13 | 575 | 11 | 255 | 9 | 408 | 14 |
| Ireland | 22,960 | 15 | 478 | 19 | 132 | 21 | 378 | 15 |
| France | 23,670 | 14 | 579 | 9 | 121 | 22 | 364 | 16 |
| Australia | 20,530 | 17 | 520 | 15 | 261 | 7 | 344 | 17 |
| Belgium | 24,630 | 12 | 502 | 16 | 180 | 16 | 315 | 18 |
| Spain | 14,960 | 20 | 418 | 24 | 91 | 23 | 312 | 19 |
| United States | 34,260 | 2 | 655 | 5 | 351 | 5 | 312 | 19 |
| Greece | 11,960 | 22 | 528 | 14 | 140 | 20 | 311 | 21 |
| Germany | 25,050 | 9 | 588 | 7 | 149 | 19 | 286 | 22 |
| New Zealand | 13,080 | 21 | 490 | 17 | 209 | 13 | 230 | 23 |
| Canada | 21,050 | 16 | 682 | 3 | 369 | 4 | 230 | 24 |
| Argentina | 7440 | 25 | 201 | 25 | 14 | 25 | 121 | 25 |

*Rank order correlations (significance levels):*
*Income with Fixed = 0.685 (0.00), with Internet = 0.507 (0.01), with Mobile = 0.318 (0.12); Fixed with Internet = 0.625 (0.00), with Mobile = 0.06 (0.76); Internet with Mobile = 0.332 (0.105).*

*Sources: International Marketing Data and Statistics, European Marketing Data and Statistics, The World Bank, and authors' calculations.*

other, none of these variables has a significant correlation with mobile phone penetration in terms of Spearman's rank correlation coefficient.

Nations vary considerably in the penetration of mobile Internet access compared to fixed-line telecommunications and fixed-line Internet access. Whereas the United States has been a global leader in overall Internet access, it lags far behind Europe and the advanced economies of the Asia-Pacific region in terms of mobile Internet access (Table 2). One reason for this is that

*Table 2: A Comparison of Fixed and Wireless Internet Users in Western Europe, the United States and the World*

|  | 2000 | 2002 | 2005 |
|---|---|---|---|
| **World** | | | |
| Internet users, million | 414 | 673 | 1174 |
| Wireless Internet users, million | 40 | 225 | 730 |
| (Wireless as proportion of Internet users) | (9.7%) | (33.4%) | (62.1%) |
| **United States** | | | |
| Internet users, million | 135 | 169 | 214 |
| Wireless Internet users, million | 2 | 18 | 83 |
| (Wireless as proportion of Internet users) | (1.5%) | (10.7%) | (38.8%) |
| **Western Europe** | | | |
| Internet users, million | 95 | 148 | 246 |
| Wireless Internet users, million | 7 | 59 | 168 |
| (Wireless as proportion of Internet users) | (7.4%) | (39.9%) | (68.3%) |

*Source: eTForecasts and authors' calculations (date: December 2002).*

fixed-line access is relatively cheap and generally not metered in the United States, whereas mobile Internet access is more expensive and largely metered. In most other developed countries, the continued prevalence of per-minute-billing of fixed-telephone-line usage makes calling plans for mobile phones comparatively attractive. The success of Japan's i-mode, beyond the indisputable technical merits of the service, is attributable in large part to the high cost and modest penetration of fixed-line Internet access. In fact, the adoption of mobile technology does not follow any single universal logic or pattern. While mobile technology in advanced nations is usually a successor or a complement to earlier generations of telecommunications, in developing parts of the world, to varying degrees, it represents an infrastructure alternative to fixed-line communications.

*Table 3: Factors Influencing the Diffusion Patterns of Mobile Technologies*

| Factor | Elaboration and Explanation |
|---|---|
| Inherent diffusion-accelerating attributes | ▪ *Mobile technologies have inherent diffusion-accelerating attributes:* Potential to save time (relative advantage), ability to connect to existing telephone network (compatibility), operation method same as the "regular" phone (low complexity), status-conferral to potential buyers (observability), and possibility to borrow a friend's cellular phone or handheld device for trial (trialability) |
| Mobile technology effects | ▪ *Many factors lower the barriers to adoption:* Low fixed and operating costs of mobile networks, ability to operate in areas with no electricity, low social barriers to adoption, infrastructure resources less prone to theft and vandalism, geographical flexibility, and innovative pricing (such as prepaid services) |
| Rapidly deployable technology | ▪ *Mobile networks can be deployed rapidly:* Ongoing reductions in fixed and operating costs due to progressively cheaper and increasingly powerful components enable rapid deployment |
| Infrastructure effects | ▪ *Large, established fixed-line networks create positive externalities:* First- and second-generation mobile phones in advanced nations benefited from such network externalities <br><br> ▪ *Relative lack of fixed-line infrastructure favors cellular networks:* Mobile networks are more attractive than fixed networks in developing countries that lack fixed-line infrastructures |
| Market size and industrial demand effects | ▪ *Diversity and size of industries affects uptake of data services:* The uptake is rapid when there are large, diverse industries likely to use mobile communications and commerce applications |
| Income and leapfrogging effects | ▪ *Income levels influence mobile penetration rates and technology generations:* High-income countries adopt early but end up having mixed generations of mobile phones, while low-income countries may adopt later with uniformly new generations of technology |
| Cultural factors | ▪ *Culture influences adoption rates and styles:* Cultural factors affect the preference for mobile phones over fixed phones; they also influence handset sizes and style preferences |
| Provider competition effects | ▪ *Competition-driven innovations by providers often unlock latent demand:* New marketing or pricing plans sometimes trigger substantial increases in use (e.g., one-rate plans in the United States) |
| Portal design factors | ▪ *Availability of reliable and user-friendly interfaces:* Uptake heavily influenced by the quality of portal design (e.g., WAP, i-mode) |
| Policy-related factors | ▪ *Government policies influence the mobile sector:* Public investment often funds backbone networks; telecom policy affects competition in and reorganization of the mobile telecom sector |

# FACTORS IMPACTING DIFFUSION PATTERNS OF MOBILE TECHNOLOGY AND M-COMMERCE

Laid out in Table 3 are the primary factors responsible for the meteoric growth rate of mobile phones and m-commerce technologies, as well as the highly heterogeneous adoption patterns just discussed. While design issues surrounding portals and interfaces doubtless influence the short-term uptake of m-commerce services (e.g., the flop of WAP versus the success of i-mode), the emphasis in Table 3 is on basic economic factors favoring the rapid growth and diffusion of mobile technologies generally.

In Table 3, insights from prior literature on the diffusion of mobile technology and m-commerce are incorporated. For example, the inherent diffusion-accelerating attributes are factors underlined by Rogers (1995, pp. 245-246), who sees mobile phones as having an "almost ideal set" of product characteristics:

- Rapidly falling cost and the potential to save time offer *relative advantage*.
- Ability to connect to existing telephone network increases *compatibility*.
- Use of the same method of operation as the "regular" phone results in *low complexity*.
- Status-conferral aspects of mobile phones boost their *observability*.
- Possibility to borrow a friend's cellular phone for trial increases the *trialability* of mobile devices.

These classic diffusion-accelerating factors help explain the exceptionally rapid rate of cellular phone adoption. Similarly, Dholakia and Kshetri (2002, 2003) argued that several "mobile technology effects" also act as spurs to mobile technology provision and adoption. These include the following:

- Low fixed and operating costs of mobile networks.
- Ability to have mobile service, even in areas with no electricity.
- Low social barriers to adoption.
- Compared to the expensive copper wiring that is lucrative for thieves, mobile infrastructure is less prone to theft and vandalism.
- Geographical flexibility, in terms of covering difficult terrain without the need to lay copper wire.
- Innovative pricing, such as prepaid service plans.

Rapid developments in mobile technology made the "mobile technology effects" more prominent and the perceived attributes of mobile phones closer to ideal. A study conducted by Yankee Group found that worldwide wireless prices fell by an average of 38% between late 1996 and early 1999 (*The Economist*, 1999). Reductions in fixed and operating costs were more dramatic in some countries. In China, for example, connection fees as well as handset prices halved from 1997 to 1999 (*The Economist*, 1999). These costs are declining further. As James Bond, the head of the World Bank's telecommunications division, pointed out:

> *(T)here is a limit to how much cheaper fixed lines can get because they involve heavy investment in labor and materials. Conversely, mobile phones share the propensity of all digital technologies to become both cheaper and more powerful over time. (The Economist, 1999)*

Furthermore, mobile sets are becoming hybrids between computers and phones. Third-generation (3G) and fourth-generation (4G) cell phones are bundling the functionalities of a phone, a computer, the Internet, and a credit card. These mobile sets allow high-speed data transmission, and the costs are likely to be lower than that of a personal computer, making the adoption more attractive for broad groups of potential users. In the Asia-Pacific region, for instance, the launch of 3G services has the potential to fuel the growth of mobile phones:

> *In the fixed-line network, it is market liberalization among the emerging giants of the [Asia-Pacific] region which is promoting growth; in the mobile network it is the launch of 3G services which promises growth, while for the Internet it is the development of more local language content which will spur growth. (ITU, 2000)*

The emergence of prepayment pricing structures, one of the "mobile technology effects" mentioned in Table 3, has been a major factor driving the diffusion of mobile telephones. In 2000, the number of prepaid subscribers of Millicom International increased by 70%, with a 79% increase in prepaid minutes (MIC, 2000). Similarly, in 1999, 70% of new users in Thailand and 100% in Malaysia were prepaid users (*The Economist*, 1999). In the Czech Republic, Eurotel Praha, a mobile service company, is overcoming the lack of a credit culture by providing prepaid services via their mobile phones for shopping online (ITC Executive Forum, 2001).

## 3G and Other Mobile Technologies

The influence of technological, political, and other environmental forces on recent m-business technologies differs significantly from the mechanisms that influenced the diffusion of earlier cellular phones. While small Nordic nations such as Sweden and Finland pioneered in mobile telephony, large (and affluent) European and Asian nations—Germany, the United Kingdom, Japan—will probably spearhead the innovation process for next-generation mobile technology and business applications. In a cross-sectional study of major economies from Asia, Europe, and Latin America, Lehrer, Dholakia, and Kshetri (2002) found that the penetration rate of fixed phones has a significant effect on the diffusion of first-generation (1G) and second-generation (2G) mobile phones but not on leading indicators of m-business technologies, especially investment in 3G infrastructures. Market size (measured by population), on the other hand, appears to correlate with strong investment in 3G infrastructures. The wide range of possible applications for 3G—advertising, business data, e-mail, information services, SMS, transactions, voice (Johansson, 2001)—appears to favor large countries with diversified industrial bases. At the same time, several factors inhibit the move to 3G and 4G technologies. These include customer behaviors anchored strongly in PC usage and easy availability of "free" or low-cost services, including voice and Internet access.

Income influences the penetration level of mobile technology as well as the optimum combination of different generations of mobile phones. High income allows potential adopters to afford higher prices while embracing an innovation (Dekimpe, Parker, & Sarvary, 2000). In an international context, it can be argued that an economy's standard of living and the level of economic development influence the adoption timing as well as diffusion speed (Antonelli, 1993; Gatignon & Robertson, 1985; Dekimpe, Parker & Sarvary, 2000; Gruber & Verboven, 2001). A certain minimum level of income is therefore a prerequisite for an effective penetration level of mobile technology. For example, 3G and 4G mobile phones are likely to be more attractive for high-income economies than for low-income economies.[4]

In addition to income, the cellular standards adopted by a country are also likely to influence its national m-commerce potential. Given the variety of mobile standards worldwide today[5] (Table 4), multiple dominant designs can be expected to coexist and compete for an extended period of time in m-

*Table 4: A Comparison of Different Standards of Mobile Phones in Use Today*

| Standard | Description | Where Used? | Remarks |
|---|---|---|---|
| *First generation (1G)* | • Based on analog technology<br>• Became available during the late 1970s and early 1980s | Worldwide | |
| Nordic Mobile Telephone (NMT) | • First commercially available analog system<br>• Introduced in 1979 | Sweden, Norway | |
| Advanced Mobile Phone Service (AMPS) | • Considered to be the "most successful" analog standard<br>• Introduced in 1982 | Worldwide | |
| Total Access Communications System (TACS) | • Based on AMPS | Originally specified for the United Kingdom | • Extended TACS (ETACS) is primarily used in Asia-Pacific countries |
| *Second generation (2G)* | • Digital wireless standards that concentrated on improving voice quality, coverage, and capacity<br>• Designed to support voice | Worldwide | |
| Global System for Mobile phone communications (GSM) | • First commercially available digital standard<br>• Relies on circuit-switched data<br>• The basic development that supports data at low rates (<9.6 kbps) has been used for e-mail from laptops | Europe, Asia | • The United States accounts for 3% of the worldwide GSM market<br>• The United States introduced GSM in 1995<br>• GSM customers estimated to reach 1.4 billion by 2005 |
| Time Division Multiple Access (TDMA) | • Introduced in 1992<br>• Also known as "North American" digital standard | North America, Latin America, Asia, Eastern Europe | • Originally, it was known as IS-54; now IS-136 (TDMA IS-136) |
| Personal Digital Communications (PDC) | • Primary digital standard in Japan | Japan | |
| IS-95 | • Based on "narrowband" CDMA technology | South Korea, North America | |

*Sources: http://intel.com/technology/itj/q22000/articles/art_6.htm, http://www.cfo.com; http://www.etsi.org/frameset/home.htm?/pressroom/Media_Kit/GSM.htm; http://www.mobileinfo.com/3G/3G_Wireless.htm; http://www.mobile3g.com/GetContracts.asp; http://www.cellular.co.za/technologies/3g/3g.htm; http://www.refreq.com/WAPTech/wap_glossary. htm; http://www.socketcom.com/pdf/TechBriefMobilePhone.pdf; Lehrer, Dholakia, & Kshetri (2002); and authors' research.*

*Table 4: A Comparison of Different Standards of Mobile Phones in Use Today (continued)*

| Standard | Description | Where Used? | Remarks |
|---|---|---|---|
| Enhanced Second Generation (2.5G) | • Builds upon the 2G standards by providing increased bit-rates and bringing limited data capability<br>• Data rates range: 57.6–171.2 kbps | Worldwide | |
| High-Speed Circuit-Switched Data (HSCSD) | • Provides access to four channels simultaneously, providing four times the bandwidth (57.6) of a standard circuit-switched data transmission rate (4.4 kbps) | North America and Europe | • Marks the first step toward 3G services<br>Predecessor of GPRS |
| D-AMPS IS-136B | • Introduced in 1999<br>• The first phase of D-AMPS provided up to 64 kbps<br>• The second phase provides up to 115 kbps in a mobile environment | North America, Latin America, Australia, and parts of Russia and Asia | |
| General Packet Radio System (GPRS) | • Supports data rates up to 171.2 kbps (about three times faster than today's fixed telecom networks and 10 times as fast as current circuit-switched data services on GSM) | Europe | • Standard from the European Telecommunicatio ns Standards Institute (ETSI) |

commerce applications. Yet, adoption of specific standards creates a trajectory of change and growth. South Korea, for instance, with its Code Division Multiple Access (CDMA) standard of 2G, will find it less costly to upgrade to the CDMA-based 3G standard than Spain (with its non-CDMA 2G standard), and will, therefore, be able to adopt 3G more rapidly. CDMA-based networks in the United States will benefit similarly.

While in Table 4 an inexorable evolution toward higher generations of mobile technology is suggested, disruptive technologies have begun to emerge. For example, the already uncertain profitability of 2.5G and 3G networks is endangered by the proliferation of so-called Wi-Fi or 802.11 hotspots. These are low-cost LANs that allow multiple wireless users within the radius of a few hundred feet to share a single broadband connection. Although Wi-Fi does not provide the blanket geographical coverage of wireless networks, it offers the speed of a fixed-line broadband connection, something that current 3G networks have no prospect of achieving. The technical and commercial disappointments of 3G networks to date raise the possibility that some countries may largely pass on 3G and one day leapfrog to 4G. In addition, the

*Table 4: A Comparison of Different Standards of Mobile Phones in Use Today (continued)*

| Standard | Description | Where Used? | Remarks |
|---|---|---|---|
| Third Generation (3G)[1] | • Will provide wide-area coverage at 384 kbps and local area coverage up to 2 Mbps<br>• Will supplement standardized 2G and 2.5G services with wideband services<br>• Will use packet switching instead of circuit switching; hence, no need to establish a continuous connection that dedicates a circuit for each call | Worldwide | Each packet contains a destination address and a sequence number so it can be independently routed and reassembled into a complete message |
| CDMA2000 | • CDMA multicarrier<br>• Also known as IS-2000. | United States | Expected to be compatible with CDMA and GSM/TDMA |
| W-CDMA | • Also known as Wideband CDMA<br>• Currently the leading 3G standard<br>• Subvariants: WCDMA-FDD used in Japan, WCDMA-TDD dominates in Europe | Europe, Canada, Japan | • In Europe, W-CDMA is known as UMTS (Universal Mobile Telephony System) |
| | • Based on AMPS | Originally specified for the United Kingdom | • Extended TACS (ETACS) is primarily used in Asia-Pacific countries |
| TDMA-SC | • Also known as TDMA Single Carrier<br>• UWC-136 (Universal Wireless Communications) and EDGE (Enhanced Data Rates for GSM Evolution) fall in this standard<br>• Will provide higher speed without changes in channel structure, frequency, or bandwidth | North America | • EDGE is a radio-based high-speed mobile data standard with aggregate transmission speeds of up to 384 kbps when all eight time slots are used |
| *Fourth Generation (4G)* | • Planned to have higher transmission rates (at least 100 Mbits/sec)<br>Technological alternatives to fixed-frequency transmission for achieving such rates include ultrawideband (UWB) transmission | Worldwide | • Originally anticipated by 2010, but Japan's NTT DoCoMo announced plans to implement by 2006, possibly using UWB |
| *Fifth Generation (5G)* | • Still in speculative phase<br>• Term used by some to refer to "infrastructureless" ultrawideband (UWB) networks that would use handsets instead of base stations to relay signals (so-called "ad hoc" network) | Worldwide | • Applications of ad hoc UWB networks to date mainly military; FCC approval for limited UWB use granted Feb. 2002 |

exorbitant cost of 3G spectrum licenses in some countries (the United Kingdom, Germany) may indirectly favor the emergence of technological alternatives to 3G because of the financial burden they impose on the companies responsible for installing 3G infrastructures.

Adoption patterns of mobile technology also depend on sociocultural factors. In newly industrialized Asian economies, for instance, people are more comfortable with smaller electronic devices and mobile handsets (Wilson, 2001). In China, on the other hand, the really small wireless phones initially did not sell well, because they were "too inconspicuous" and did not offer adequate opportunity to show off (low "observability" attribute). Later, the public attitude toward these wireless devices changed in China as well. Among the world's English-speaking countries, commonalities of language and close cultural links appear destined to elevate the United Kingdom to a "lead market" for 3G applications (Lehrer, Dholakia, & Kshetri, 2002).

As with other technologies, policy-related factors play an important role in the diffusion of mobile technology and m-commerce applications. For instance, government investment in the telecom sector, government initiatives to encourage mobile phone purchases, and intense competition in and reorganization of the telecom sector are found to be the major causes of China's rapid mobile telecom network growth (Kshetri & Cheung, 2002). The experiences of countries such as South Korea and Sri Lanka confirm the general rule that competition among mobile operators leads to lower prices and rapid mobile network growth (UNDP, 2001).

# A FRAMEWORK OF
# FUTURE DEVELOPMENT

The future direction of mobile commerce development is surrounded by considerable economic and technological uncertainty. As the recent history of PCs and the Internet showed, the growth and success of IT *hardware* depends critically on the availability of attractive *software* packages. Such packages range from specific software applications (so-called "killer apps") to portal designs that facilitate the user's overall interface with the technology. Particularly in developing economies, the development of the m-commerce landscape requires the emergence of country-specific mobile portals with specialized sites for Web phones. The evolution of m-commerce in such economies has been hampered to date by the lack of such portals. In developed countries, on the other hand, a frequent lament is that the key "killer application" for m-commerce has yet to be identified, again hampering penetration of the new technology.

With the caveat of such uncertainties in mind, summarized in the framework shown in Table 5 are some of the major regional differences that emerge from comparative analysis. Turning first to the vertical dimension (the global centrality of nations in developing m-commerce applications), a dichotomy between core (active developers) and peripheral (passive adopters) nations seems likely. As shown in the upper-right corner of Table 5, countries in the economically most-developed regions (North America, Western Europe, Japan) possess ideal supply and demand conditions for pioneering m-commerce applications. On the supply side, they possess the financial and technological resources to invest in high-risk ventures. On the demand side, they possess the high levels of income and sophisticated industrial users to afford and demand novel applications.

Nonetheless, the considerations explored in our prior analysis give some reason to expect an "expanded core" of pioneering nations in m-commerce, including, in particular, East Asia (upper-left corner of Table 5). This is due, in part, to the mobile Internet handicap of the United States. Traditionally the world's leader in IT innovation, the United States is now saddled with such extensive sunk investments in fiber cables that investment in mobile Internet infrastructure (particularly in 3G) is lagging. The reasons are more general than just the U.S. lag, however. These reasons are captured in the horizontal dimension of Table 5, the distinction between countries where mobile telephony supplements fixed lines and countries where it genuinely substitutes for fixed lines. In the latter, especially certain developing countries in Asia and South America, mobile penetration is much higher than Internet access. Can this be turned to an advantage for business and development through m-commerce (ITC Executive Forum, 2001)? Particularly in the East Asian "Tiger" countries (Taiwan, Hong Kong, South Korea, and Singapore), with their specialization in consumer electronics and computing products, there is good reason to think that it can.

Recent contracts awarded by some developing Asian economies to telecom vendors in the United States and Europe indicate that many of the advances in wireless communications are likely to be installed in Asia before in North America and Europe. Fiber optic and cable networks enjoyed by consumers in the United States simply do not exist in most Asian countries. Furthermore, contrary to the popular belief that the wireless Internet will only

*Table 5: Comparative Regional Environments for M-Commerce Development*

| Global Centrality in M-Commerce Innovation | Infrastructure Role of Mobile Telephony | |
| --- | --- | --- |
| | *Substitute for Fixed Lines* | *Supplement to Fixed Lines* |
| *Active developer* | **Industrialized East Asia** (e.g., Japan, South Korea, Taiwan) | **North America, Europe** (e.g., the United States, Scandinavia) |
| *Passive adopter* | **Developing Countries** (e.g., Argentina, Sri Lanka) | **"Null Set"** (but may be filled for 4G) |

*Source: Authors' research.*

be successful in markets with lower PC penetration, economies with high PC penetration, such as South Korea, are also showing phenomenal success in m-commerce.[7] For example, Korea Telecom Freetel launched 1X technology[8] at the end of 2001 and achieved a 35% penetration rate of wireless Internet use within five months, with 1X users consuming twice as much airtime as other mobile users (Luna, 2002). Furthermore, after introducing Binary Runtime Environment for Wireless (BREW)[9] applications, data revenue increased by 60%.

Demand and supply factors influence the potential of East Asian countries to become "lead markets" for m-commerce innovation. The use of wireless communications as a substitute for fixed-line access to the telephone and Internet network creates a demand advantage to the extent that it translates into earlier and higher diffusion rates in the population's use of the new technology than in other advanced countries. Just as important, however, are supply-side factors. Like Japan, East Asian Tiger countries feature a large number of producers specialized in innovative, high-technology products for global IT markets. Such East Asian countries, therefore, have the means to translate feedback from behavior in their domestic markets into novel product designs. Having already climbed the ladder from technology imitators to active technology developers, these countries have the requisite understanding of global markets to pioneer innovative m-commerce products, especially in components and hardware.

In contrast, not all developing countries using wireless telephony as a substitute for fixed lines can be expected to be global innovators in the use of the new technology. Developing countries using mobile networks as a cost-effective alternative to fixed lines in large rural areas (Sommermeyer, 2001), for example, will not necessarily be major innovators in using and developing the technology. In the context of the Asia-Pacific region, analyst Frank Yu of Ion Global argued that developing markets like China do not have a need to broaden their existing wireless penetration base by adding more services. He argued: "They have other basic needs they want to work on before they start experimenting with prototype systems. They'll just let Japan and Korea do that" (quoted in Stout, 2001). As the earlier example of GWCom illustrates, however, this fact does not preclude countries like China from pioneering innovative uses of existing wireless technologies. Meanwhile, the proximity of the huge Chinese market may provide advantages to Japanese and Korean producers in their exploration of new m-commerce applications.

# DISCUSSION AND CONCLUSION

Mobile technology, mobile Internet access, and m-commerce are growing rapidly on the global stage; however, growth rates vary widely across economic regions. The penetration rate of mobile phones, the optimum combination of different generations of telecommunications, and the combination of different technical formats vary according to a wide array of economic, sociocultural, and policy-related factors. The ways in which companies integrate m-commerce applications into their business models depend upon numerous environmental factors, particularly the combination of communications technologies previously adopted and the mobile technologies currently diffusing in their domestic economies. Given the scope for technological leapfrogging and alternative national mixes of fixed lines and wireless infrastructures, global heterogeneity in national patterns of m-commerce development—and hence, of business models in m-commerce—appears to be a likely prospect for the foreseeable near future.

An interesting consequence of this global heterogeneity is the tenuousness of the distinction between "leading" and "lagging" countries. Nations that "lag" in fixed-line telephony or Internet use stand to benefit proportionately more from mobile technologies and may, in some cases, be better positioned to innovate their use; though it is always dangerous to speculate, there is some indication that industrialized East Asian countries may fall into this category. Yet not all nations that rank as "leaders" in m-commerce necessarily have cause to celebrate; many European countries, for example, are concerned that their carriers exhausted their financial resources in the development of 3G and may be unable to follow through in infrastructural development of mobile technologies. By the same token, the U.S. "lag" in m-commerce may not prove to be a long-term handicap; in fact, the U.S. lag in 3G may accelerate the exploration of technological alternatives like Wi-Fi or UWB. Thanks to NTT DoCoMo's i-mode, Japan is assuredly a "leader" in m-commerce; yet this leadership position is partly conditioned by the fact that i-mode represents, to a significant extent, a Japan-specific substitute for fixed-line Internet usage; the industrial and export benefits of this leadership position are far from certain.

# ENDNOTES

[1] To simplify usage, the term "e-commerce" is employed for electronic commerce transactions carried out via fixed, wired terminals, and the term "m-commerce" is for electronic commerce transactions carried out via mobile, wireless terminals. In the larger sense, both are variations of electronic transactions, but using the prefix "e" for fixed/wired and "m" for mobile makes it easier to contrast these two types of electronic transactions.

[2] See http://www.gwcom.com/html/news0303.htm.

[3] The examples in the foregoing paragraph are of using mobile phones as aids for regular commerce. The step to m-commerce, however, is a short one, when reliable and easy-to-use data-ready phones become widely available.

[4] Leapfrogging effects of developing countries adopting the latest mobile technologies are going to taper off, as the developing countries accumulate sizable segments of mobile users. With 3G, and especially with 4G, only the poorest and most backward developing nations are apt to engage in leapfrogging. In other developing nations, the "mixed generation" pattern of the affluent economies will begin to take hold.

[5] A variety of standards can be found even in a continent. For example, at the end of the first quarter of 2001, Latin America had 32 million TDMA subscribers, 15.9 million users had CDMA handsets, and 4.9 million had GSM handsets (Petrazzini & Hilbert, 2001).

[6]   The standards mentioned here are from the recent IMT-2000 (International Mobile Telecommunications-2000) recommendation.

[7]   PC penetration in South Korea is about 50%, similar to levels in the United States (Luna, 2002).

[8]   1X (or CDMA2000 1X) is the name that identifies the 3G technology that upgrades CDMA networks and wireless devices to 3G features and services.

[9]   BREW platform, designed by QUALCOMM, is a thin application execution environment providing an open, standard platform for wireless devices. Carriers' BREW-based services enable wireless users to customize their handsets by downloading applications over the air from a carrier's application download server.

# REFERENCES

Antonelli, C. (1993). Investment and adoption in advanced telecommunications. *Journal of Economic Behavior and Organizations*, 20(February), 227-245.

Banks, C. (2001). The third generation of wireless communications: The intersection of policy, technology, and popular culture. *Law and Policy in International Business*, Washington, (Spring).

Bradley, S. P. & Sandoval, M. (2002). Case study: NTT DoCoMo—The future of the wireless Internet? *Journal of Interactive Marketing*, 16(Spring), 80-96.

Chen, P. (2000). Broadvision delivers new frontier for E-commerce. *M-commerce,* (October 25).

de Haan, A. (2000). The Internet goes wireless. *EAI Journal*, (April), 62-63.

Dekimpe, M. G., Parker P. M., & Sarvary, M. (1998). Staged estimation of international diffusion models: An application to global cellular telephone adoption. *Technological Forecasting and Social Change*, 57(January/February), 105-132.

Dekimpe, M. G., Parker P. M., & Sarvary, M. (2000). Global diffusion of technological innovations: A coupled-hazard approach. *Journal of Marketing Research*, XXXVII(February), 47-59.

Dholakia, N. & Kshetri, N. (2002). The global digital divide and mobile business models: Identifying viable patterns of e-development. In S. Krishna & S. Madon (Eds.), *Proceedings of the Seventh IFIP WG9.4 Conference* (May 29-31, pp. 528-540). Bangalore, India.

Dholakia, N. & Kshetri, N. (2003). Mobile commerce as a solution to the global digital divide: Selected cases of e-development. In S. Krishna & S. Madon (Eds.), *ICT and Development*. London: Ashgate Publications.

Dholakia, R. & Dholakia, N. (2003). Mobility and markets: Emerging outlines of M-Commerce. *Journal of Business Research*, (forthcoming).

Ebusinessforum.com. (2000). *GW Trade: Serving a high-tech niche in China*. Retrieved March 23 at: http://www.ebusinessforum.com.

*The Economist*. (1999). *Survey: Telecommunications*, (October 9).

Gatignon, H. & Robertson, T. S. (1985). A propositional inventory for new diffusion research. *Journal of Consumer Research*, 11(March), 849-867.

Gruber, H. & Verboven, F. (2001). The diffusion of mobile telecommunications services in the European Union. *European Economic Review*, 45, 677-588.

ITC Executive Forum. (2001). *E-Commerce to M-Commerce: The next stage in the digital revolution*. Available at: http://www.intracen.org/execforum/docs/ef2000/eb20008.htm.

ITU. (1997). *World telecommunication development report* 1996/97: Trade in telecommunications, International Telecommunications Union, 1997. Available at: http://www.itu.int/plweb-cgi/fastweb?getdoc +view1+www+27994+7++waiting%20time%20and%20telephone.

ITU. (2000). Digital mobile growth, International Telecommunications Union.

ITU. (2000). *ITU Telecom Asia closes: Region's mobile and Internet boom set to continue*. Available at: http://www.itu.int/plweb-cgi/fastweb? getdoc+view1+www+38110+3++china%20and%20mobile.

Johanssen, P. (2001). The future is here. *Pacific Telecommunications Review*, 22(3), 21-31.

Kshetri, N. & Cheung, M. (2002). What factors are driving China's mobile diffusion? *Electronic Markets—International Journal of Electronic Commerce and Business Media*, 12(1), 22-26.

Lehrer, M., Dholakia, N., & Kshetri, N. (2002). National sources of leadership in 3G m-business applications: A framework and evidence from three global regions. *Proceedings of the Americas' Conference on Information Systems* (AMCIS), Dallas, Texas, (August 8-11).

Lindsay, G. (2000). An answer in search of a question: Who wants m-commerce? *Fortune*, 142(9), 398-400.

Lopez, A. (2000). The South goes mobile. *UNESCO Courier*, (July/August). Available at: http://www.unesco.org/courier/2000_07/uk/connex.htm.

Luna, L. (2002, May 27). Mobile data gets real. *Telephony*, 242(21), 38-41.

MIC. (2000). *Millicom International Cellular S.A. Annual Reports and Accounts.*

Petrazzini, B. A. & Hilbert, M. (2001). *3G mobile policy: The case of China & Venezuela. International Telecommunications Union.* Available at: http://www.itu.int/osg/spu/ni/3G/casestudies/chile-venezuela/Chile-Venezuela.PDF.

Rai, S. (2000). In rural India, a passage to wirelessness. *New York Times*, (August 4), C1-C3.

Rogers, E. M. (1995). *The Diffusion of Innovations* (fourth ed.). New York: Free Press.

Secker, M. (2001). Does m-commerce know where it's going? *Telecommunications*, 35(4), 85-88.

Shaffer, R. (2000). M-Commerce: Online selling's wireless future. *Fortune*, (July 10).

Sommermeyer, M. (2001). *Asia: Telecom's next frontier.* Available at: http://www.hightechcareers.com/docs197/asia197.html.

Stout, K. L. (2001). China Mobile has eyes only for 2.5G. *CNN.Com*, (June 5).

UNDP (United Nations Development Program). (2001). *Human Development Report 2000.* New York: UNDP. Available at: http://www.undp.org/hdr2001/completenew.pdf.

Wai, T. C. (2001). *WAPping into developing countries.* Retrieved July 10, 2001 at: http://www.sibexlink.com.my/g15magazine/g15mag_vol3apr_infotech.html.

Wellenius, B., Braga, C. A. P., & Qiang, C. Z. (2000). Investment and growth of the information infrastructure: Summary results of a global survey. *Telecommunications Policy Online*, 24, 8/9.

Wilson, D. (2001, January 15). Is Asia ready for M-Commerce? *Electronic News*, 47(3).

Wooldridge, A. (1999). Survey telecommunications: At the back of beyond. *The Economist,* (October 9).

Zuckerman, A. (2000). International technology trends. *Transportation & Distribution*, 41(9), 64-68.

**Chapter II**

# Assessing the Market Potential of Network-Enabled 3G M-Business Services

Mats Samuelsson, M-B Networks, USA

Nikhilesh Dholakia, University of Rhode Island, USA

## ABSTRACT

*Argued in this chapter is that the key differentiators of m-business comprise a set of the experience/function parameters that set m-offerings apart from e-business. "Network-enabling" of m-business by taking advantage of new network-based services that can seamlessly handle many of the service features can add great value to m-business offerings. In network-enabled m-business, business adopters and their end users in the field are freed from the burden of constantly dealing with the challenges of designing, redesigning, configuring, integrating, upgrading, troubleshooting, maintaining, and billing for m-service offerings. A constellation of m-business service offerings that could take advantage of new network-based services is presented. Finally, an approach is suggested to add value and differentiate m-service offerings so that they continue to remain profitable.*

# INTRODUCTION

While with the early introduction of third-generation (3G) wireless networks, some financial and technical hindrances to the launch of 3G/Universal Mobile Telecommunication System (UMTS)[1] networks and services were seen, by 2010, mobile business (m-business) services are likely to be pervasive in the economically advanced countries of the world. An emergent class of new, network-enabled services will drive a substantial part of the growth in 3G/UMTS business applications. Even newer generations of 4G services may enter the scene by 2005. In this chapter, an approach for distinguishing, conceptualizing, and evaluating the market potential of such network-enabled 3G services is presented.

The chapter is divided into seven parts. The part following this introduction describes the multiple dimensions on which m-business applications, especially in the business-to-business (B2B) arena, offer advantages and enhancements over regular, computer-based electronic commerce (e-commerce). Next, the concept of "network-enabled" m-business services is explained. This is followed by profiles of selected, proposed, new UMTS/3G m-business services in the B2B arena. Using an evaluative framework that we developed, we offer a comparative assessment of the profiled m-business services. Finally, we offer general guidelines for assessing new m-business services, discuss some of the challenges in introducing them, and provide a summary and some key conclusions.

# M-BUSINESS: ADVANTAGES AND ENHANCEMENTS OVER E-BUSINESS

E-business has conquered the world. Despite the bursting of the dot.com bubble, it is hard to believe today how one managed to transact any business in the early 1990s without the Internet. Whether employed for information, support, or advertising, nearly every business in the world of any size has a Web site. E-commerce has revolutionized how many companies do business, allowing for new business models and spawning completely new types of businesses. So with e-business less than 10 years old, is the world ready for something new, something with the potential to revolutionize business practices the way e-business did? The answer is "Yes."

Like e-business that preceded it, m-business as a transformational force is here to stay. In the next few years, m-business will emerge as a powerful new approach for conducting business. The initial years will be spent solving some of the challenges associated with the new m-business paradigm—usage and user patterns, creating optimal terminal types, and evolving sustainable business models. But after this is accomplished, m-business will become as pervasive as e-business has become today. While the transformation induced by m-business would be dramatic, it would not necessarily replace e-business. M-business would enhance existing e-business functions and applications and launch new ones, totally mobile instead of being tied to desktop terminals. In many ways, m-business would establish new patterns of doing electronic transactions, over and beyond what fixed-line e-business is capable of.

E-business happened because of the combined efforts of the personal computer (PC), telecommunications, business software, and office technology industries. M-business, similarly, will happen because of the combined efforts of the world's mobile handset manufacturing, telecommunications, computers, software, and office technology industries. The Internet and e-business made the PC industry grow to a level where PC makers were shipping 100 million PC units every year by 2002. Even without a significant m-business base, by 2002, the mobile handset makers were already shipping 400 million units every year. In this massive global business, m-business is appearing as a new platform for creating product and service differentiation. Internet and e-business helped drive the supply and demand for multimedia computers. The underlying chip and display technology required for m-business is in its infancy, with the first "primitive" (by future standards) multimedia/application-capable terminals introduced in 2002. As m-business matures, it would transform the handset, rendering it as different from its predecessors as today's desktop PC screen is from the green-tinted, nongraphic PC screen of the early 1980s.

So what do we understand by m-business and by m-commerce? Parallels are drawn between e-business/e-commerce and m-business/m-commerce in Table 1. The descriptions in Table 1 are generalized and may give the impression that the only thing m-business does is replace the PC-based Web access with mobile phone/handheld terminal-based access. The Wireless Application Protocol (WAP), commonly viewed as a failure (at least in its initial version), provided a Web browser optimized for mobile phone and handheld terminal use. Among many reasons for the initial troubles of WAP was that it force-fit desktop-oriented screen content into small terminals. If m-business were merely an extension of the desktop e-business into handheld devices, then its role would be limited.

The analogies in Table 1, derived from the perspective of today's e-business, are inherently limiting. Under this frame of thinking, m-business would simply extend e-business to the mobile terminal. This fails to take into account that the "m" is fundamentally different from the "e". A simple look at the vibrant consumer market for cell phones provides some clues. In this dynamic marketplace—visible especially in the electronic retail districts of cities like Tokyo, Hong Kong, and Singapore—hundreds of colorful models compete for the consumer (and business persons' attention), with new models brought out every month. The PC, on the other hand, basically looks like it has for the past 20 years, with only one or two dramatic innovations—usually introduced by

*Table 1: Basic, but Inherently Limiting, Distinctions Between E-Business/ E-Commerce and M-Business/M-Commerce*

| E-Commerce | M-Commerce |
|---|---|
| Web-based solutions for selling and interacting with customers | Mobile phone/handheld terminal-based solutions for selling and interacting with customers |
| | Services delivered to customers via mobile phones/handheld terminals |

| E-Business | M-Business |
|---|---|
| Web-based extensions of the business enterprise | Mobile extensions of the business enterprise |
| Web- and computer-based solutions for improving business productivity and performance | Mobile phone/handheld terminal-based solutions for improving business productivity and performance |
| Web- and computer-based solutions that allow for the launch of new business models | Mobile phone/handheld terminal-based solutions that allow for the launch of new business models |

*Source: Authors' research.*

*Figure 1: M-Business and E-Business Distinctions Viewed in an Experience/*
*Function Framework*

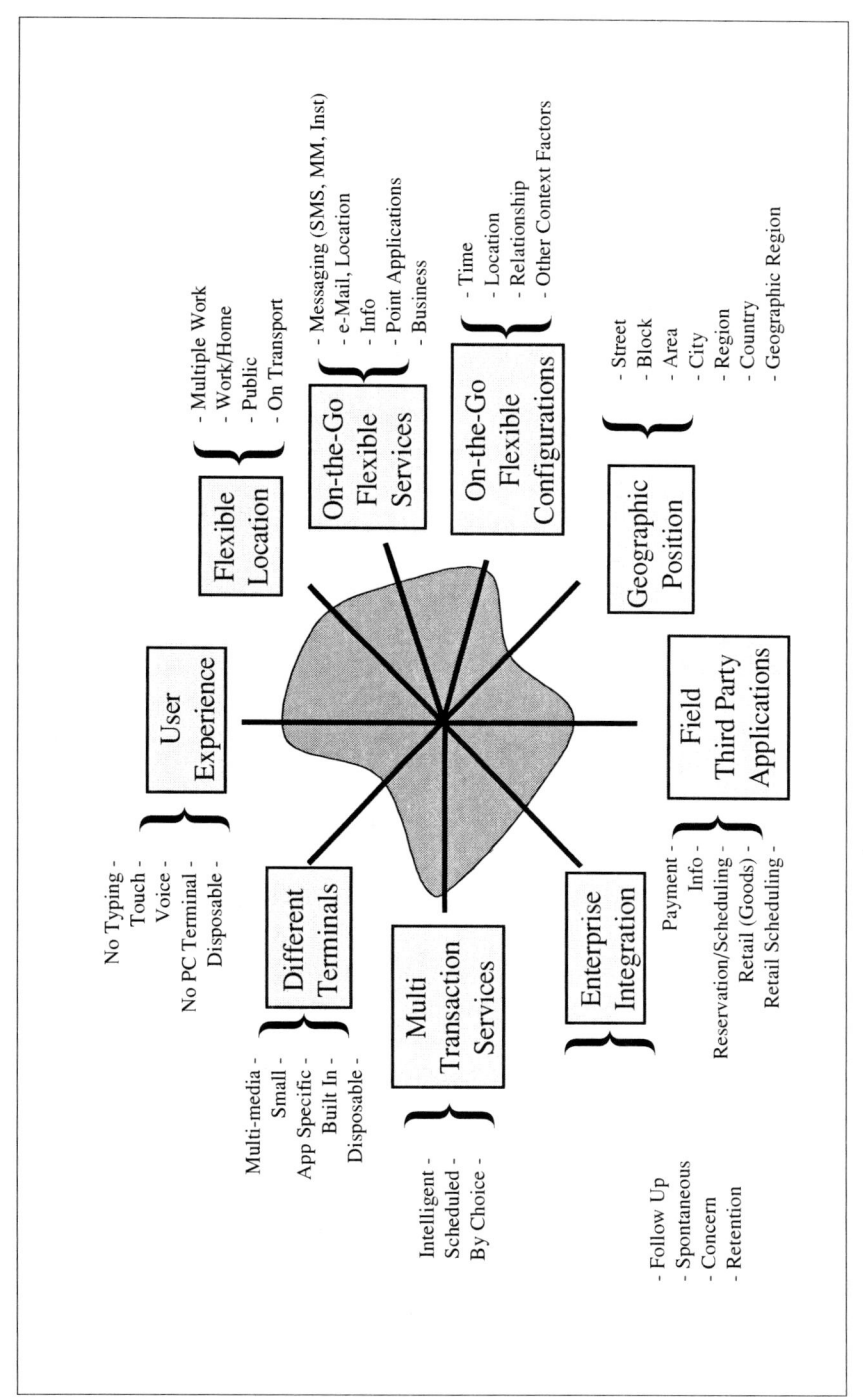

*Source: Authors' research.*

Apple Computers—along the way. A look at Figure 1 hints why this is the case. Out of the 10 experience/function variables that m-business will draw its power from, e-business (and e-mail) can deliver at best only four or five, and some of those are delivered with severe limitations.

M-business's ability to draw on all of these experience/function variables (and some that have not yet been imagined) is the key to its revolutionary power. The best way to describe this is to say that m-business solutions 10 years from now will be as dramatically different from today's terminal and e-business applications as today's multimedia PC is from the teletype interfaces to mainframe or minicomputers of the 1970s.

In the following paragraphs, we provide short descriptions of each of the 10 experience/function variables that are likely to set m-business apart from e-business.

## User Experience

Perhaps the biggest differentiator between e-business and m-business is the sensory experience of the user. In e-business, the user is in a stationary position in front of a PC terminal and interfaces the content using a keyboard and point-and-click devices. In m-business, this is replaced by total mobility, and the terminal can be voice or touch activated.

## Different Terminals

A disposable terminal is probably the most radical way of describing how different terminals could be. This is not a farfetched idea—in 1990, no one would have thought of a disposable camera. Today's manufacturing technology aided by the unrelenting progress of Moore's Law[2] will allow an ever-increasing differentiation of terminal offerings. Terminals that are bendable, so that they can be rolled up, have been demonstrated at trade shows. Miniature-sized terminals allow for packaging into ever-changing shapes and forms. Prepaid phone service is just the introduction to other prepaid services, complete with "free" terminals. Multimedia is here to stay and will continue to evolve.

## Multitransaction Services

M-business services could be scheduled and delivered in multiple ways. Users can choose to have a variety of services delivered at the times and places that they specify. In some cases, the services can be prescheduled (for peak hours, late night, birthdays, etc.). In still other cases, the network and the device can make intelligent assessments of what services are needed and then proffer such services.

## Integration with Enterprise Applications

With m-business, a business enterprise could move most of its capabilities out into the field. Services and applications that required office visits and meetings could now be delivered on the go, with full access to all enterprise applications residing on business IT and information systems.

## Field Third-Party Applications

Terminals that are m-commerce ready can receive services not just from the primary wireless service providers but also from a variety of third-party providers. Most of these third-party providers would work through the wireless service operators. In some cases, the terminal may be able to communicate directly to third-party wireless service providers, through ad-hoc information exchange setups or direct connectivity. The source of applications and information therefore becomes transparent to the user.

## Geographic Positioning

From a continent to the corner of a street, m-commerce networks would be able to locate the user and tailor the service mix to the geographical location, keeping in view the constraints and opportunities of the geographical setting as well as the preferences of the user. A service would therefore work differently in Singapore than in Hong Kong, London, or New York based on profiles or regional preferences.

## On-the-Go Flexible Configurations

Today's user profiles, whether in e-business or m-business settings, show the way to flexible configurations. But rather than requiring manual setups and changes, the m-services of the future will be automatically configured. So the minute a user leaves the home area, the service will be automatically configured with ring-tones, forwarding information and even downloaded information as the user travels. And should the user want to configure it in a new way, a simple code will download a new configuration. This would be the world of hundreds of prepackaged user experiences ready to be activated.

### Integration with Mobile Services

New m-business services would be easy to integrate with preexisting mobile services. For example, m-business offerings could easily incorporate a variety of existing messaging services, short message service (SMS), and e-mail. They could also use conference bridges, network-based calling, voice mail, as well as many emerging services, like downloadable handset applications, multimedia messaging, and information services.

### On-the-Go Flexible Services

With easier integration of services, users would be able to avail of prepackaged as well as programmable service mixes. Some m-business systems would offer a service palette from which the users would be able to choose and blend a variety of services.

### Flexible Location

With m-business, the user can work, do daily chores, and play, at work, home, recreational, shopping, and vehicular locations. The coming blurring of roles in the era of m-business will spawn multiple opportunities as well as trigger major social changes.

## NETWORK-ENABLED M-BUSINESS SERVICES

Extrapolating existing business approaches and paradigms into new areas is the most obvious way of looking into the future. After all, we are comfortable with what we see today and can easily see how it can be used tomorrow. For m-business, the problem with this approach—treating m-business as a simple extrapolation of e-business—is that it fails to take into account the dramatic differences (as well as different capabilities) between the two. Some of the most dramatic differences are screen size[3] and the mobile user experience. But, equally important is the fact that m-business services will be built (assembled) from different "piece-parts" than e-business. Wireless service operators will deliver some of these "new" piece-parts, and many of these are being discussed and implemented today. Examples include location information Application Programming Interfaces (APIs) and services. Needless to say, there will be other, as-yet-unknown service piece-parts.

There are obviously many m-business applications developed and deployed today. All major overnight package delivery services utilize specialized terminals and have Web pages on which customers can track delivery progress. These solutions are developed directly for companies by systems integrators or IT suppliers. They are typically limited to using wireless services already offered by operators, although the size of some of these companies means that they are able to get special services developed and deployed by service operators. There will obviously continue to be a large market for customer-specific data services that utilize wireless transport and messaging services. There is also a substantial standardization under way to develop and deploy network services APIs that allow third parties to tap into and directly interface with network services. This is clearly a first step toward network enabling of data services.

Network enabling, however, can mean much more than just offering access to network-based data services and APIs for location, messaging, etc. A useful way to look at network enabling is in terms of successive categories associated with implementation or business focus. Shown in Figure 2 is a progression of m-business value addition, as the service offering moves from mere data transport to additional services to APIs for network services, and finally to complete integrated service solutions. It is at the right-hand extreme of Figure 2 that the true revolutionary potential of m-business becomes evident. In the following paragraphs, we describe the progressively higher stages of value addition and service integration.

## Basic Data Transport Services

At the most basic level, adding data transport capabilities to simple mobile voice telephony opens some opportunities for m-business. The evolution of Web-browsing from today's slow WAP speeds to higher data rates will revitalize some of this market. Pure data transport to support custom terminal-based network applications, like those used by today's package delivery services, will continue to grow as enterprises start to capitalize on higher-speed data transport to develop new business productivity and enhancement applications.

## Additional Network Services

Enhancing basic mobile data access and Web-browsing capabilities with additional network services and specialized terminals adds more value to the m-business concept. Examples of this include the Blackberry handheld device, which provided mobile e-mail and messaging capabilities. Finland's Benefon provides GPS capabilities in the mobile phone, making it a useful device for navigating in cities as well as in the wilderness. Multimedia messaging is certainly positioning itself as a major value-added service, replacing today's SMS as a key data service. Some network data services will utilize location information, for example, delivering messages only in certain areas.

## APIs for Network Services

The APIs for network services allow for tighter service integration of messaging, location-based services, usage monitoring, and billing. These APIs

*Figure 2: A Progressive Model of M-Business Value Addition*

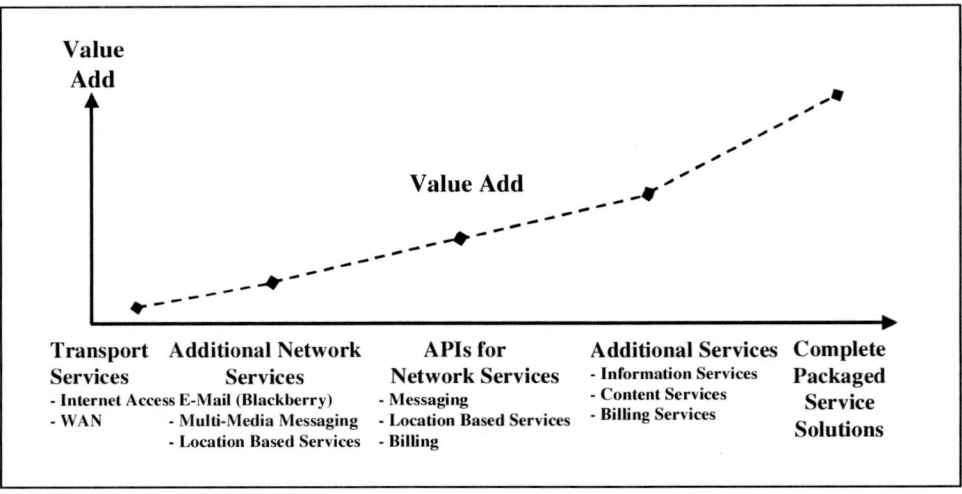

*Source: Authors' research.*

are intended to be used by third parties or business enterprise applications to offer services that are more closely integrated with network services, utilize network billing, or deliver services that are based on where the user is located.

## Additional Services

As an additional value-adding step, the wireless operator can offer additional m-business-oriented services providing complete value-added information, tracking, billing, or messaging services. Business customers can utilize these complete service packages in order to develop more complete applications for their users.

## Complete Integrated Service Packages

Finally, as an ultimate value-adding step, the m-business service provider can design and offer fully integrated service packages that solve complete problems. In the next main section of this chapter, we illustrate and assess such service packages.

## Crystallizing the Concept of Network Enabling

Together, these five areas of service "add-ons" represent the concept of network enabling, i.e., by integrating services with capabilities inherent in the network or offered by the network service providers, services become enabled by these.

Looked at in a slightly different way, they can be viewed as "value-adds" to the core "transport function" of the wireless network. They also represent a migration of the business that wireless operators are striving for. The wireless service operators want to move from basic "telephone" service to more value-added services. Network enabling, the gradual enhancement of service offerings, with network capabilities, is an important implementation aspect of this move. M-business services can, of course, be implemented in a simple fashion by using only transport services from wireless operators. Such services are in use today and are basically terminal applications that communicate with the business enterprise IT systems. But the future evolution of m-business services will be more network enabled by adding and integrating various value-added network services to the way they operate and are implemented.

*Figure 3: Illustrative Constellation of M-Business Solution Concepts*

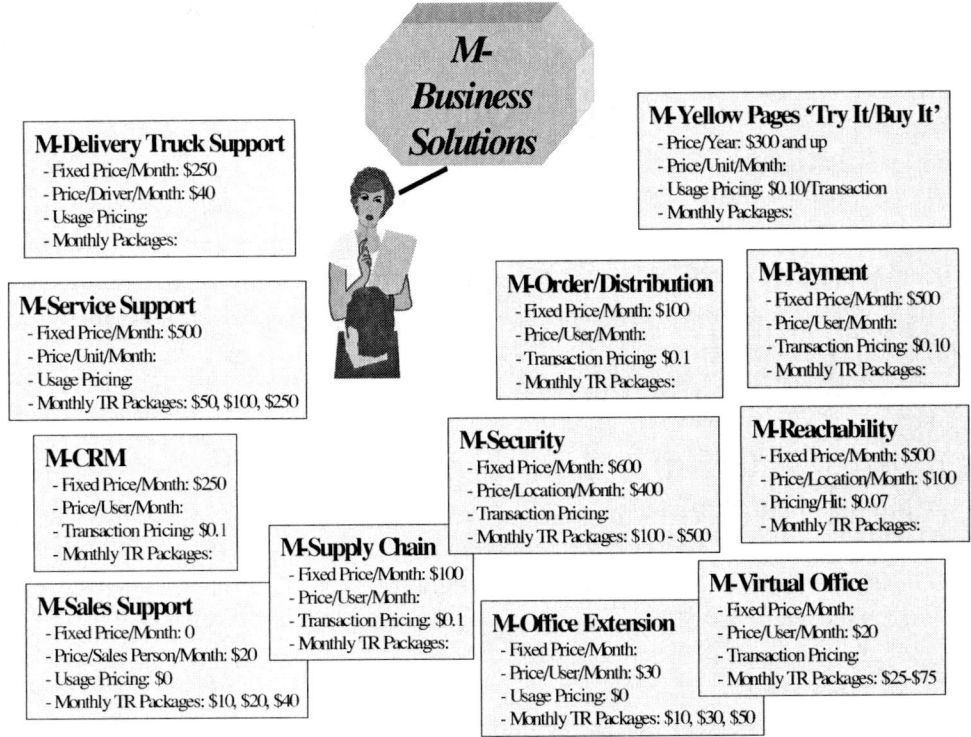

**M-Business Solutions**

**M-Delivery Truck Support**
- Fixed Price/Month: $250
- Price/Driver/Month: $40
- Usage Pricing:
- Monthly Packages:

**M-Yellow Pages 'Try It/Buy It'**
- Price/Year: $300 and up
- Price/Unit/Month:
- Usage Pricing: $0.10/Transaction
- Monthly Packages:

**M-Service Support**
- Fixed Price/Month: $500
- Price/Unit/Month:
- Usage Pricing:
- Monthly TR Packages: $50, $100, $250

**M-Order/Distribution**
- Fixed Price/Month: $100
- Price/User/Month:
- Transaction Pricing: $0.1
- Monthly TR Packages:

**M-Payment**
- Fixed Price/Month: $500
- Price/User/Month:
- Transaction Pricing: $0.10
- Monthly TR Packages:

**M-CRM**
- Fixed Price/Month: $250
- Price/User/Month:
- Transaction Pricing: $0.1
- Monthly TR Packages:

**M-Security**
- Fixed Price/Month: $600
- Price/Location/Month: $400
- Transaction Pricing:
- Monthly TR Packages: $100 - $500

**M-Reachability**
- Fixed Price/Month: $500
- Price/Location/Month: $100
- Pricing/Hit: $0.07
- Monthly TR Packages:

**M-Supply Chain**
- Fixed Price/Month: $100
- Price/User/Month:
- Transaction Pricing: $0.1
- Monthly TR Packages:

**M-Sales Support**
- Fixed Price/Month: 0
- Price/Sales Person/Month: $20
- Usage Pricing: $0
- Monthly TR Packages: $10, $20, $40

**M-Office Extension**
- Fixed Price/Month:
- Price/User/Month: $30
- Usage Pricing: $0
- Monthly TR Packages: $10, $30, $50

**M-Virtual Office**
- Fixed Price/Month:
- Price/User/Month: $20
- Transaction Pricing:
- Monthly TR Packages: $25-$75

*Source: Authors' research.*

# PROFILES OF SELECTED, PROPOSED, NETWORK-ENABLED 3G/UMTS M-BUSINESS SERVICES

The illustrative m-business service concepts in Figure 3 are intended as profiles of real service solutions that will become available to business and consumer users in the years ahead. In all cases, the business services capitalize on the network-enabled business services infrastructure provided by 3G/UMTS operators or service providers. This infrastructure is connected to the business customer's IT environment through standards-based interfaces for communication, information transfer, and messaging. It also utilizes network information, thereby providing the business customer with enhanced information from terminals, input devices, locations, usage patterns, as well as the ability to interface with third-party systems and third-party service providers. In the following paragraphs, each of the concepts in Figure 3 is elaborated upon.

## M-Delivery Truck Support

This service is a delivery support service designed to increase the effectiveness and productivity in medium and small delivery fleets. It can also be targeted at larger delivery fleets in cases where the fleet operators do not have their own internal implementation, or to modernize older existing bulkier terminal and application implementation. The service provides drivers with one terminal-based application that follows them throughout the day as they make deliveries. Information about each customer and delivery is downloaded as the delivery route progresses. Customers can be called by pressing a button (phone number downloaded), directions can be obtained by pushing another button (map and directions downloaded), and detailed information for each customer can be downloaded prior to each delivery. Customers can be asked questions at the time of delivery, with answers entered using application-assigned buttons. Such questions can include special promotions, offers, or customer satisfaction feedback. Dispatch operation knows where each driver is (location-based service) and can easily contact the drivers via messaging for changes in delivery or special pickups. Finally, the terminal can only be used for specified applications and phone numbers, eliminating unauthorized use.

## M-Yellow Pages—"Try It/Buy It"

The m-business version of *The Yellow Pages* revolutionizes the service by adding total content flexibility and query-specific information (code and location dependent), allowing each response to a Yellow-Pages request to be customized.

The service works via "operator-branded" logos with an alphanumerical code located as part of a display (in-store, billboard, etc.) or in advertising (sign, magazine, etc.). The user enters this code into a simple Yellow-Pages terminal application. This starts a Yellow-Pages query service that connects to the right supplier's IT site. This site contains Yellow Pages information (company, address, phone, product information, special offers, etc.) that can be dependent on the following:

- Time: Different information during different times of the day.
- Date: Changing information dependent on date.
- Query site: Information dependent on where Yellow-Pages request originates.
- Query code: Single company can have multiple codes dependent on product, service, site, etc.

## M-Sales Force Support/Communication/Management Application

This application is a general field sales support m-business service that allows a mobile sales force to tie directly into their company's IT support structure with a mobile device of choice. Designed to work with various types of mobile computing devices (and even cell phones), the application creates a user interface for the business situation at hand, whether downloading product information, creating orders and delivery schedules, delivering sales campaigns, or obtaining customer location information.

In a network-based m-business implementation, the terminal application and its content drive the service function. For example, by choosing a product line button, the salesperson will be connected to different office systems at different locations (and even different companies), depending on the product line. The network also automatically routes messages and voice calls to the right sales support functions, such as installation, scheduling, etc. The installation message not only contains the message information, but also which network service logic the response should communicate with. The network service logic routes the response to a third system or company, together with a notification forwarded to the original salesperson (and maybe a formal communication note to their end customer).

Many applications will consist of interfacing to the business customer's IT environment [through standards-based Internet connectivity—Web-service, extensible markup language (XML), messaging, etc.]. Other application implementations will utilize services from third-party providers or customize device applications depending on specific device applications or input commands. The result is a General Packet Radio Service (GPRS)/3G terminal device application customized for the business customer's need, dependent on specified content interfaces to their business IT infrastructure, rather than requiring the business customer to develop and integrate a complete business application.

## M-Service Support Management and Information Application

An application similar to the previous one could focus on service force management and provide customers with support information through mobile devices. As described above, this application is integrated into the customers' business IT and business process environments, adding mobility. Features such as message content-driven service logic applications and third-party information and content services could be added.

An advantage of this application is the ability to interact with all the systems, services, and support functions required for speedy completion of a service call. This communication starts while the service representative is on the way to the job. The connectivity is driven by business-specific service logic that interprets messages sent to the home office. So instead of initiating a number of separate communications with IT systems and support staff, the service representative can handle the whole service transaction as one application. First, the GPRS/3G terminal is connected to the system being serviced, and then it relays information collected to the right supplier product support site for diagnostics. The support site can, for example, recognize a need for software upgrade, sending a message to a third-party system software upgrade site authorizing the distribution of a software download. Through network service logic, this software gets distributed to the right GPRS/3G terminal (or end customer's IT system) at the same time as the network service initiates a service call transaction starting the creation of a billing transaction in the home management information system (MIS). At the same time, a signature application is initiated in the GPRS/3G terminal, where the end customer can sign off on the service call. The service force management application then assigns the next service call to the service representative based on his or her position and time. As it does this, it places a call from the GPRS/3G terminal to the next customer to advise them about technician arrival time. Again, this application shows how network-based service process logic can coordinate multisystem/multicompany transactions using messaging, voice, and XML.

## Mobile Office Solution for Small- to Medium-Size Businesses

The Mobile Office Solution provides total mobility for the business, including messaging and integration into business-specific applications and business processes. This advanced integration is done through the XML/Web services extension to the wireless device. Advanced messaging integrates with the office messaging system, and a number of potential add-on applications allow for mobile meetings, note sharing, and application/time/location-based information and alerts. Mobile GPRS/3G devices can vary from today's laptops and PDAs to future specialized terminals and applications.

The advantage of the Network Enhanced Office Solution is that it can provide different support, depending on where the user is and what terminals they use. A scenario illustrates the possibilities:

> *When the user exits the office, the network automatically detects this and starts an automatic process of communications between the office IT environment and the GPRS/3G terminal that the user carries. This interaction will be different for PDAs, portable computers, ordinary cell phones, or the new 3G phones. Instead of having to download specific information by hand before leaving, this is automatically done for the user, in the right format for the right device. This filtering is then applied to the user requests. For example, all messages can be held until later in the day or when the user returns to the vicinity of the office. High-importance messages are delivered in the right format to applications that can interpret responses (such as set-up call in five minutes). Based on this information, the network then creates a call or information interchange at the appropriate time, automatically. And, as third-party services are needed, they are automatically provided to the user. There is no need for individual subscriptions and billing transactions: as part of the Mobile Office service, the service operator handles such details.*

The M-Office solution truly becomes a mobile office where person/ terminal/home office/interaction/mobility become one instead of just being a series of wireless messages and phone calls going back and forth. The partial mobility implementations prevalent today are replaced by tomorrow's seamless mobile office.

## Mobile CRM Solutions for Small/Medium Size Businesses

The mobile customer relationship management (CRM) application provides a flexible approach for each business customer to handle customer relationship management. By providing CRM support to employees and to the end customers' mobile environments, a small- or medium-size business can dramatically improve customer service.

This application allows a business to increase responsiveness and offer truly mobile CRM, a solution that follows the end customer and delivers appropriate timely information when, where, and in the form that the customer wants it. It is again the concept of replacing a large number of point-to-point communications with a content- and process-driven service. Instead of making a phone call or surfing through Web sites to straighten out a billing problem, the GPRS/3G terminal application defines the issue and then forwards this information to the supplier. Based on the nature of the issue, the supplier's CRM application defines a customer satisfaction response entailing multiple transactions. These are forwarded to the mobile CRM (m-CRM) transaction application and set in motion the following transactions and processes:

1.   Immediate message to customer acknowledging receipt of complaint.
2.   Message communication to third-party service providers to correct the problem.
3.   Correction message is forwarded to the customer when responses are received, and an appropriate response and explanation is downloaded directly to the customer's terminal; if there is a financial transaction, this is also simultaneously downloaded to the customer's terminal (or bank account or mobile cash).
4.   Personal phone call is placed (when customer is available, based on availability information entered from the GPRS/3G terminal) apologizing for mistake (can also be a friendly video message); a token gift (a glass of wine) is offered as an apology.
5.   Download of the gift transaction to the customer's favorite wine-bar (or one close by, based on mobility information), where the customer gets their wine by simply telling their name; bartender would already have received name and message when the customer walked in the door, based on location information.

With the network-based m-CRM service process logic, it is possible for an end user to set corrective action in motion through a CRM application on their GPRS/3G terminal. Once the event has been triggered and forwarded to the business, the enterprise's IT environment interacts with the network business process logic and communicates (using XML messages) with appropriate suppliers, functions, personnel, and service and help desks, plus communicates the appropriate relationship-cementing information to the appropriate places (such as the wine bar).

## M-Supply Chain Management for Small- to Medium-Size Businesses

The supply chain management application introduces wireless supply management solutions to small- and medium-sized enterprises. It adds smart mobility, transactions monitoring, and communication device transparency, plus the ability for a company's ordering and supply systems to interact via network-enhanced XML messages. The standard XML messages fit seamlessly within a Web services supply chain IT application environment, while XML fields controlling network services interact with these and other resources (network or third-party applications) via XML-SN applications processes.

M-supply chain management (m-SCM) can mean that a terminal application, rather than extensive (and expensive) IT applications, becomes the collecting point for supply information delivered by systems that were not integrated. This is accomplished by specifying the XML-enhanced fields that tie into and control the applications.

M-SCM also provides integrated views into all suppliers' ordering and shipment systems, without having to procure and integrate specific expensive software applications for the task. Users can also receive the exact information that they need, wherever they are, on the terminal of their choice.

With m-SCM, messaging is integrated with real-time voice and data communications and can be used to streamline supply processes.

It implies access to enhanced procurement services that are offered as part of the service by third-party service providers specializing in their respective fields. Because the third-party providers are fully integrated into the network, these services are provided seamlessly to the user.

Also, m-SCM creates a platform for integrated ordering and shipping solutions, without the need for integration. The application comes together on the mobile terminal.

## M-Security and Alert Generation Application

The business security application integrates a GPRS/3G terminal security application with the business security procedures of small- to medium-sized companies. Whether there is a need for interaction with a live security person or automated voice, pattern, or password protection, this application extends company facility and data security to mobile security devices, allowing the business to reduce costs and increase security reliability. Some features included in such a service are as follows:

- Remote video monitoring (from any GPRS/3G video-capable terminal).
- Voice print recognition.
- Entry of numerical codes.
- Alert messages that are context dependent, i.e., based on time, day, sequence of events, and circumstances.
- Third-party monitoring and alarm applications.

## Retail Order/Distribution Application

Using wireless service providers and customized terminals, this application provides the same wireless business order/delivery applications to small-to mid-sized companies that large corporations implemented through extensive integration efforts of enterprise IT environments. Viewed as a sort of captive "FedEx" for every business, this application provides interface capabilities similar to those used by suppliers and customers in order distribution systems, replacing custom terminals with GPRS/3G JAVA applications. The network-based service flow applications provide not only the order distribution connectivity but also the ability for suppliers, distributors, and customers to have the right information on their handheld devices at the right time. By specifying a general-purpose XML interface for this application, each part of the supply chain can implement their information interface on an XML Web site, thereby simplifying integration with their respective IT order management and billing systems.

## M-Reachability and Location Communication Application

The m-reachability application provides the business user with a way to communicate with their customers based on proximity, service, or promotional agreement. As the user walks down the street or drives along the road, the terminal m-reachability application displays information from businesses along the way on the user's terminal, as each business establishment comes within walking or stopping range.

In addition, a business location or event (such as a sports event, art fair, street festival, etc.) will now know how many of their customers or "fans" are nearby and will be able to communicate with them instantly, when the potential fans are in the vicinity of the event. This application provides microbroadcasting capabilities from a location specific to its selected target audience within specified driving or walking ranges.

## Electronic Payment Application

Electronic payment applications have slowly emerged in the GSM world. GPRS/3G adds a location/service process dimension, as well as integration with third-party payments and applications providers. Products and services can now suddenly be paid for without dependence on traditional credit card companies or banks. The network version of this brings the application to the masses, giving every small- and mid-sized business the ability to create electronic payment business models without the need for Web browsers or Internet business models.

This application can be as easy as creating a mobile "cash register," employing a third-party payment application, while the end customer is standing in front of the checkout clerk. Card sliding and passwords are replaced by a screen message and OK button on the customer's own GPRS/3G terminal. Alternatively, advertising is augmented by a code (or barcode) that instantly brings all necessary information to the terminal, where simple button clicks set in motion messages and interactions between the customer and the small business, free from expensive overhead infrastructures. Services and messages of the following type can be provided:

- Where can I get it—Map to closest location.
- Have it ready for me—Sure.
- Send it to this address—Sure.

## The "Officeless" Business Applications

This business application provides a mobile location-less office environment, complete with computing, messaging, and business process functions. The only difference is that such an "office" is virtual, i.e., not associated with a physical business address or facility. The "officeless" back office is housed in a secure unmanned hosting location, and all LAN and PBX functions and connectivity are provided by the network. Connectivity within the officeless business is provided with 3G, managed by an officeless business process application capable of integrating all 3G terminal-based applications with back office and communications.

By eliminating the need for office and administrative space and replacing it with 3G terminal applications and a business process communication/information application, the service provider can offer a cost-saving solution that will be the foundation for new small- and mid-sized business models. Virtual offices could become virtual businesses.

As a network application, the network manages all connectivity and information exchanges through permanent virtual connections to all users' 3G terminals. "Sending e-mail" and "Dialing Phone Calls" is replaced by instant application, voice, and face-to-face communication between all participants. New or temporary employees simply connect into the officeless environment, participating in the communication and information transfer needed for their work.

## In Sum: Advantages of Network-Enabled M-Business Solutions

Network-enhanced 3GPRS/3G applications are a way to create new types of service implementations that are closely integrated with the wireless network. Such applications replace the end-to-end applications (terminal communicating with customer's IT system) with 3G service processes, wherein the terminal communicates with the service operators' business applications, that in its turn manage the communications to multiple IT systems, third parties, and other application providers.

The technologies making such services possible are being developed today (XML, Web services, high-performance service process engines, smart IP switching, XML-based UMTS applications). What is lacking is the right network service element that can integrate the parts and combine them with service development and operational tools, allowing for large network deployment. Innovative efforts to develop such network elements are currently underway.

# COMPARATIVE ASSESSMENT OF PROFILED APPLICATIONS

Comparing the "service profiles" of Figure 3 can be like comparing apples and oranges. Many such services exist in various rudimentary or specialized forms today but will obviously evolve over time. The initial phase of business-oriented wireless services can be characterized as the early attempts of an industry to address business needs. Hence, the focus is on existing niche needs, like e-mail, information downloads, or generic business user needs, like flight schedules, weather, stock market updates, or communications with the home office. The step from these rudimentary applications to the next generation of streamlined business- and user-specific applications will be slow at first but would accelerate as business users realize the operational and competitive advantages of such services. Service suppliers (wireless operators) would also begin to see the opportunities associated with addressing their business customers' needs by providing a "business solution" focus, eventually targeting corporate and individual business functions. As the new services evolve, the full range of market-seeking supply-demand variables would come into play. The result would be a mix of services, marketing approaches, and price-performance points. Some services will be priced low and others high, depending on the specific market characteristics, competitive situation, and specific customer segment variables. A "service value" framework is critical in assessing how individual services are positioned and targeted (and how the service mixes and how implementations change over time).

Such a framework is described in Figure 4, focusing on the core dimensions of "service value-add" and "service differentiation." The service value-add is comparable to the "network value-add" described while profiling the various services in the previous section. Service differentiation introduces a new and fundamental marketing aspect to the assessment. The opportunities presented by the network-aided "mobility" aspect, described in the preceding section, create new ways for differentiating service offerings over time. This

*Figure 4: Core Dimensions for Comparing Network-Enabled M-Services*

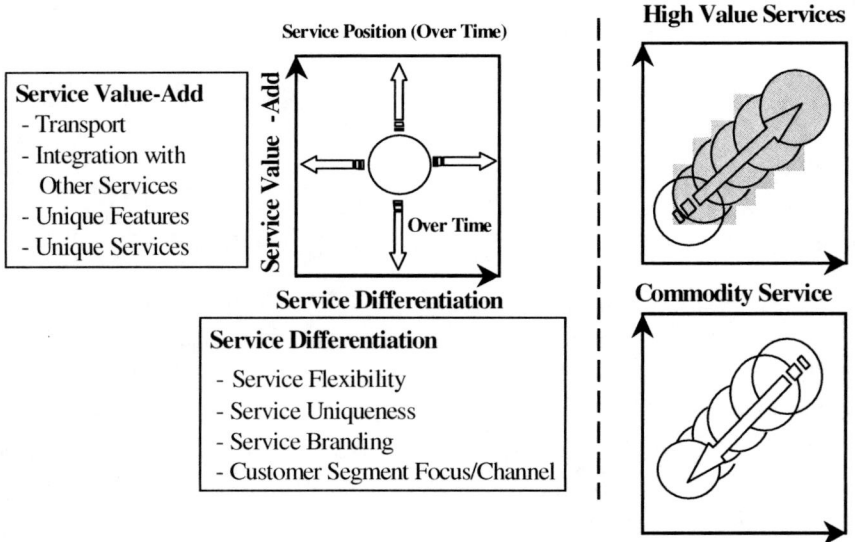

*Source: Authors' research.*

variable is critical for the m-business services marketplace; it is what makes the m-business offerings unique. For example, a simple m-Yellow-Pages service may be implemented and priced in a totally different way for a year-round delivery through a business terminal than for a onetime delivery through a disposable consumer terminal. The service may be limited in both implementations: the first focusing on B2B Yellow Pages and the second on information from, for instance, a specific retail chain that gives away the disposable terminals to be used by their customers while visiting the store.

At the same time, real service value-add often tends to be reflected in whether a service becomes a commodity service or a fully differentiated service over time. Existing mobile services tend to be a poor guide in this evaluation. They represent the results of addressing yesterday's opportunities instead of tomorrow's opportunities. The values are therefore skewed, overemphasizing existing business models instead of evolving business models. Consider, for example, the ability to imagine Web-based document distribution in the year 1990. No one today would print and distribute binders of documentation instead of just providing them online, but in 1990, this "business model" would have made little sense. In the evolving m-business models, information and effectiveness/efficiency improvements would be critical in deciding (high) value, while more mundane information distribution services will become "commodities." If implemented along these lines, high-value services have the opportunity to become high-profit services, while broadly defined commodity services would tend to be under constant price pressure over time. This is where service differentiation and value-add both come in. In order to continue to position the service in the upper right-hand corner of Figure 4, the service provider must be able to continuously add new service value-adds and provide differentiation. Unless supported by a strong external differentiator or truly unique value-adds, services tend to drift toward the lower-left corner over time.

# GENERAL GUIDELINES FOR ASSESSING NEW M-BUSINESS SERVICES

The market potential for new m-business services is driven by the volume of users that would use and buy a service and the pricing associated with service usage. Most service development starts with a simple market analysis: How many users will be using these services? The answer to this depends on how

*Figure 5: Enhancing M-Business Offerings via Experience/Function Improvements*

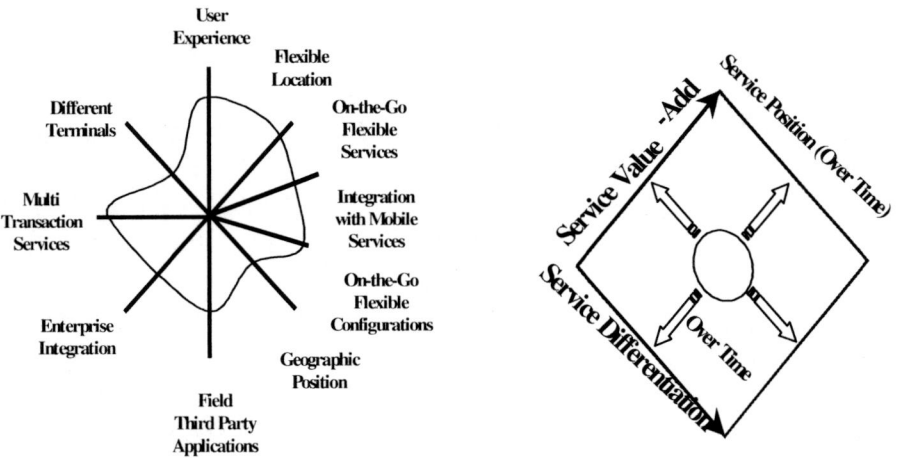

*Source: Authors' research.*

widely a service is defined. A truck delivery service will have a much smaller addressable market, but the value proposition for this service is a real productivity enhancer, perhaps increasing the productivity of a delivery fleet by 10%. Service pricing, therefore, could be relatively high on a per-user basis. When first introduced in the early 1990s, SMS on mobile phones represented a small market, but by enhancing the services with a lot of experience/function variables (most having to do with user interfaces and a simple way of coding messages), the user volume grew dramatically to today's position of SMS as the major data service in wireless networks, driven by price elasticity and user behavioral patterns.

Service value-add and service differentiation play crucial roles in determining what the true potential of a service is over time. The ability to differentiate services increases their market penetration, while innovative service value-adds can keep service pricing high. So, what factors drive the value-add and differentiation dimensions of Figure 5? In looking at many past service examples and assessing some new ones, the experience/function parameters described earlier can be used to provide service value-add or service differentiation over time. The challenge for the designers and marketers of new m-business offerings is to augment the experience/function parameters that are

*Figure 6: Scenarios for Rising and Sinking ARPU Potentials for M-Business Offerings*

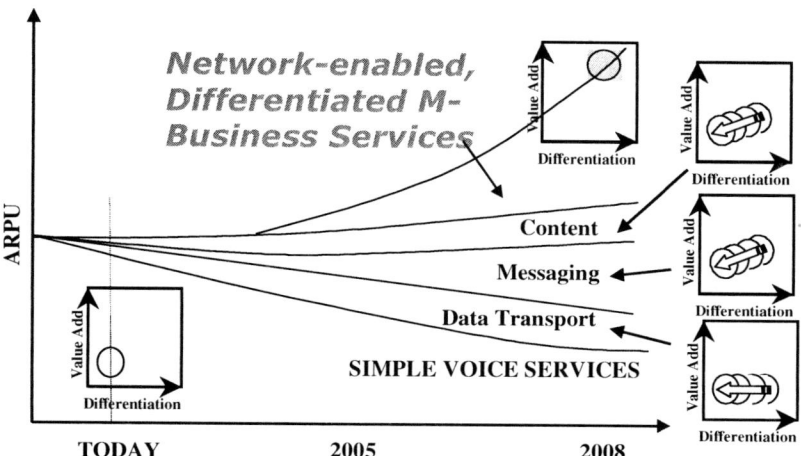

*Note: ARPU stands for Average Revenue Per User.*
*Source: Authors' research.*

most appealing to the specified target segments and that can be engineered into the designs in cost-effective ways. If such steps are taken, then m-business offerings with potentially rising average revenue per user (ARPU) can be developed. New m-businesses that fail to take advantage of the experience/ function parameters discussed in this chapter, are likely to face declining ARPU scenarios and sink into "commodity traps," where aggressive (and usually ruinous, from a profitability perspective) pricing is the only option left (see Figure 6).

When carefully designed for "value-add" and "differentiation," m-business services offer a substantial opportunity to create services that can be improved, evolved, and profitably marketed for extended periods of time. In a world where commoditization is a threat to revenues and profitability, this offers an opportunity to the GPRS/3G wireless operator to defend service offerings by evolving and enhancing them with new features and add-on services. Well-crafted m-business services designed for business users and consumers may just be the answer to the telecom operators' dilemma of declining revenues per user.

# SUMMARY AND CONCLUSION

We outlined the key differentiators of m-business, the experience/function parameters that set m-offerings apart from e-business.

For true value adding, m-business offerings should also take advantage of new m-business network services that can seamlessly handle many of the service features, thus freeing the business adopter and the end user in the field from the burden of constantly dealing with the challenges of designing, redesigning, configuring, integrating, upgrading, troubleshooting, maintaining, and billing for m-service offerings. This is what will create the true value of m-business for 3G/UMTS wireless operators. For business users, we presented a number of m-business service offerings that take advantage of new wireless network capabilities (based on new and evolving network infrastructures). Finally, we offered an approach to add value and differentiate m-service offerings so that they continue to remain profitable businesses. This requires a concerted effort on the part of service providers to understand their customers' business needs and the ability of the providers to offer an evolving m-business solution for these needs. It is difficult to foretell what exact service mixes would take off. Service providers, for example, may end up giving free 3G terminals/handsets to business customers (and maybe even to the customer's customers) in order to support highly profitable services. The alternative to developing high-value network-enabled services may be sinking into "the commodity trap," where ever lower prices rule, and profitability virtually evaporates.

# ENDNOTES

[1]    In this chapter, we will use the acronyms 3G, UMTS, and 3G/UMTS interchangeably. These denote the new class of 3G mobile telecommunications and m-commerce products, applications, and services that were launched in the early 2000s.

[2]    This refers to the exponential decline in costs, and therefore prices, of information technology (IT) as a result of advancements in technological platforms and manufacturing methods. The name is derived from Gordon Moore, cofounder of Intel, who first commented on this phenomenon with respect to semiconductors.

[3]    While it is unlikely that people will walk around with 17" display terminals, it is likely that large-format, flat, and foldable mobile terminals would appear as replacements for paper copies of newspapers and magazines.

**Chapter III**

# Mobile E-Commerce and the Wireless Worldwide Web: Strategic Perspectives on the Internet's Emerging Model

Mahesh S. Raisinghani, University of Dallas, USA

## ABSTRACT

*Telecommunications, the Internet, and mobile computing are merging their technologies to form a new business called mobile commerce or the wireless Internet. As more companies make their Web sites accessible to wireless devices and offer services and software to support wireless connections, they urgently need to consider how well the existing infrastructure will support their applications. A well-thought-out strategy is key to any major network change, especially a wireless one, which can touch many different aspects of a business organization. Provided in this chapter is an overall description of mobile commerce, the Internet's emerging model, and examination of its current state-of-the-art, opportunities and challenges, and future trends. It explains the requirements for operating mobile commerce and the numerous ways of providing this wireless Internet business. While the Internet is already a valuable form of business that already transformed the way the world is doing business, it is about to change again. This change is being driven by consumer demand for wireless devices and the desire to be connected to information and data available through the Internet. There are many new opportunities that have only begun to be explored, and this will become a large revenue source for those who capitalize upon this new form of technology. However, like other capital ventures, these new opportunities have their drawbacks, which may limit growth of the mobile commerce market if not dealt with. Mobile electronic commerce technology is changing our world of business, just as the Internet has changed business today.*

# INTRODUCTION

Mobile commerce (m-commerce) is the delivery of electronic commerce (e-commerce) capabilities directly into the consumer's hands via wireless technology and the placement of a retail outlet into the customer's hands anywhere. This form of e-commerce allows businesses to reach consumers directly, regardless of their location. The term m-commerce is a variation of the e-commerce term used for business being done over the Internet. M-commerce gained momentum in Europe and now is reaching the United States. Internet commerce presently requires the consumer to be attached to a desk with a computer terminal to conduct transactions. M-commerce allows the consumer to purchase, do banking, download cash or tickets, etc., when you need to, wherever you are, simply by using a mobile phone. M-commerce is a part of e-commerce—it is just more user-friendly than current systems. In most countries, the penetration of mobile phones is much larger than the penetration of Internet access (Duffey, 1998).

M-commerce is the integration of technologies using wireless devices for conducting business over the Internet. M-commerce can be done by computer solutions, such as laptops and palmpads, with wireless devices attached to connect to the Internet, or by using newly adapted cellular phones to receive digital transmissions of Internet material to these phones. These are all linked by software and service providers that provide the platform on which to conduct these operations.

A new business model is emerging: the integration of wireless networks with data communications, combined with e-commerce, to create wireless e-commerce.

Wireless e-commerce will generate significant revenues within the next several years from such services as wireless banking, wireless stock trading, and a variety of wireless-based shopping ventures. Wireless communications and e-commerce are already multibillion dollar global businesses. The integration of mobile communications with e-commerce already started. For years, companies in the vertical markets, such as field repair, have been utilizing mobile communications networks to enable their technicians to order parts and check inventories. The opportunities for wireless e-commerce in the horizontal markets, such as traveling executives, and the consumer market are generating much appeal (Reiter, 1999).

M-commerce is a quantum leap of technology applications and will not be limited simply to banking and brokerages. Other market uses will emerge. Payment options are one example being tested now, in which products in a store may be scanned as one walks out and automatically deducted from a Smart Card that stores cash on it from your local bank. Airline and rail connections will be enhanced with ticket reservation and payment facilities. Mobile phone users will also have access to new online auction houses to submit bids and check developments by use of the cellular phone (Brokat, 2000).

# CURRENT STATE OF M-COMMERCE

The mobile market penetration in the United States is around 41%, far less than countries like Finland with 75%, Hong Kong with 74%, or the United Kingdom with 74% but still an indication of the market acceptance of wireless communications services (Beal et al., 2001, p. 6). However, in our country, only 6% of users use their mobile phones to access the Internet, and this is a much lower percentage compared with other countries like Japan with 72%, Germany with 16%, or the United Kingdom with 10%. The percentage of people who received wireless messages or offers for products in the United States is 20%, 40% in Japan, and 25% in Germany and the United Kingdom (Beal et al., 2001, p. 9).

In Japan, estimates of current Internet users are between 19.37 million and 27 million as of February 2000. Japan's mobile Internet access increased from 3 million subscribers to 7 million subscribers in just the first five months of 2000. Their total mobile phone subscription rate is at 57 million users as of March 2000. Japan also has 10% of the world's Internet users (Eurotechnology, 2000).

*Figure 1: Japan's M-Commerce[4]*

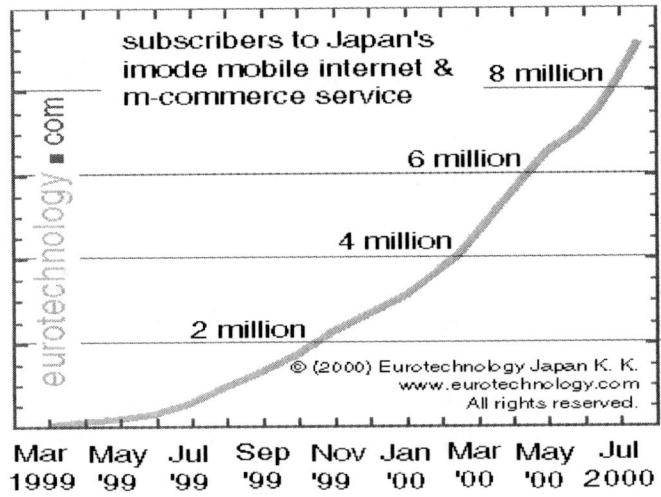

With the increasing use of small portable computers, wireless networks, and satellites, a trend to support computing on the move emerged. The worldwide number of cellular, GSM (Global System for Mobile Communications), and digital subscribers increased from 140 million in 1996 to more than 300 million in 1999 and is expected to grow to 650 million by 2001. In the United States, capital investment increased from $6.3 billion in 1990 to $66.8 billion in 1999, and service revenues were up from $4.5 billion to $38.7 billion in 1999 (Varshney & Vetter, 2000). The International Data Corporation (IDC; Framingham, Massachusetts, USA), forecast revenue generated from m-commerce to explode from an estimated $29 million in 2000 to $21 billion in 2004. As wireless applications develop, so will market demand. Jupiter Media Metrix (Melville, New York, USA) predicts the number of U.S. wireless Web users to grow from 4.1 million in 2000 to 96 million in 2005.

Wireless network operators are currently upgrading their second-generation (2G) networks with HSCSD (high-speed circuit-switched data) and 2.5G technologies such as General Packet Radio Service (GPRS) to be able to handle voice and high-volume data. HSCSD is a multi-global system for mobile communications (GSM) technology, which currently offers up to 43.2 kbps across most of Europe. A 2.5G is typically an "always-on" technology, offering basic digital mobile voice telephony with higher-speed data capabilities of up to 56 kbps. Simultaneously, network operators around the world are building

*Figure 2: World Internet Usage[4]*

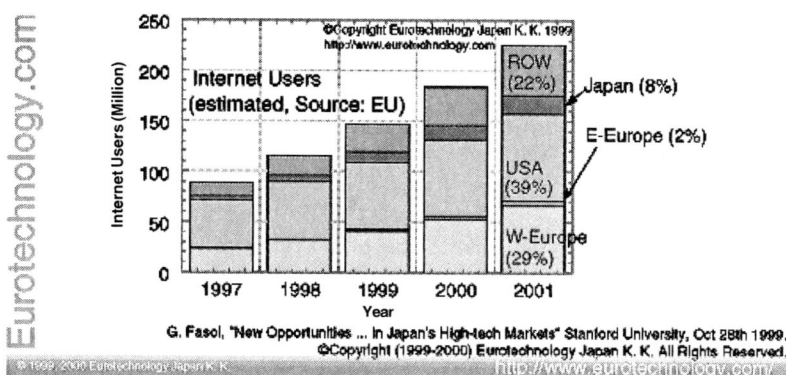

third-generation (3G—the collective third-generation wireless technologies are referred to as 3G services) networks, which offer always-on broadband networking with high-speed data access. Ultimately, 3G networks are expected to include capabilities and features such as enhanced multimedia, high speed (upwards of 2 Mbps), and roaming capabilities throughout Europe, Japan, and North America. Please see current m-commerce statistics listed in Appendix B.

# CONDITIONS SPARKING DEVELOPMENT OF MOBILE COMMERCE

A large reason for the high level of acceptance for m-commerce is the large number of mobile phone users and Internet users around the world. The Internet promoted electronic services, and m-commerce is another means of using the Internet. Customers now want to take advantage of Internet services from mobile end devices so that they can conduct business from any location in the world. The boundary between mobile telecommunications and the Internet is becoming more indistinct by the day (Brokat, 2000).

The highly lucrative industry of Internet commerce and mobile communication is a driving force in bringing many companies to develop this technology. The second catalyst is that many mobile phone users, especially in Europe and then in South East Asia, will be using smart cards, and this technology is another way to use cellular phones for business. If mobile phone operators can add functionality or convenience to these smart card applications, they will be able to create powerful new distribution channels and generate additional business (Duffey, 1997).

Much of today's wireless e-commerce technology is a result of technology being developed by many of the mobile phone makers. The Europeans led this charge, because they have some of the highest numbers of cellular phone users. This is a result of the global economic and political environment during the 1980s that promoted greater unification and collaboration, which helped the new telecommunications industries in Europe to flourish. As the need for better communication facilities grew, due to increased trade and investment flows, the solutions provided by the new technological developments become more viable (Muller & Schnoring, 1995).

# EQUIPMENT USED IN M-COMMERCE

This is an overview of the equipment necessary to conduct m-commerce. It consists of digital cellular phones, smart cards, laptop or palmpads, and the software to operate and communicate with this hardware.

## Digital Cellular Phones

Dual-slot mobile phone technology was developed in Europe. Two key things are required for it to work: phones with the capability and chip-based cards. Dual-slot mobile phones offer a suite of value-added services, including mobile banking. It can be reprogrammed in the field and has substantial free memory for further applications. This allows subscribers to turn their mobile phones into tools to support their business and leisure lifestyle (Brokat, 2000).

To understand digital cellular technology, we must understand the background of the cellular phones as it relates to speed of data transmissions. Analog technology is considered to be the first generation of cellular technologies. The second generation of digital cellular technology is high-speed circuit-switched and packet-switched data technology. High-speed circuit-switched data tech-

*Picture 1: Dual Slot Mobile Phone[3]*

nology uses a single voice channel and delivers data at a rate of 9.6 kbps. Packet-switched means the computer that is connected to the cell phone sends and receives bursts, or packets, across the radio channel. The channel is occupied only for the duration of the data transmission instead of continuous transmissions, making it more efficient than circuit-switched. The third generation of digital cellular technology refers to much higher data transmission speed in the range of 14.4 Mbps. It will enable wireless multimedia applications, such as videoconferencing (Intel, 1999).

New mobile handsets were designed with larger screens and browsers to allow users to review and select Internet content. This content is provided by a wireless markup language (WML) displayed by the microbrowser, much like hypertext markup language (HTML) is displayed on your PC. The handset features a "roll and click" device, which is like a mouse, and a small graphical user interface (Durlacher, 1998).

## Smart Cards

Smart cards are a cross between an ID card and an electronic wallet. They can be used to store and exchange money from banks as well as support the payment functions of digital cellular phones. Already used in parts of Europe, this card provides many attributes that will enable technology to better serve consumers. Some of the present mobile communication in Europe relies on dual smart card technology. It consists of one smart card, internal to the cell phone, and one external card, which can hold personal information and be used as a cash card or electronic wallet, as well as a phone card (Rundgren, 1999b).

Smart cards are plastic ID cards containing integrated circuit chips that are capable of reading, writing, storing, and processing information. The size and

*Picture 2: Smart Card*[3]

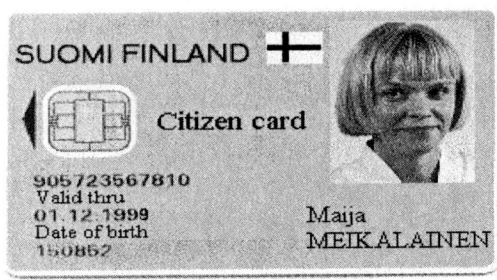

shape of the plastic, the positioning of the chip, and its resilience to attack are defined by international standards. They cost between $2 and $20, depending on their capabilities. Multiple applications include contactless smart cards that can be read by radio signal from a card reader.

Smart cards have won the battle with magnetic-strip cards because of their security, reliability, capability, and lifetime cost. The contactless smart card ticketing solution is much cheaper in the long run. Capital investment can be 90% lower, revenues can be increased by 5% to 10% through lower fraud, and maintenance can be 30 times lower with a contactless smart card system. Smart cards are already used for public transport and parking services in cities in Europe and Asia. The most important advantages of smart cards are the capability and the security that they offer. Smart cards can hold large amounts of data and do so securely (Duffey, 1997).

One advantage of smart card IDs is that they are extremely hard to forge. A personal identification number (PIN) code is added as an extra security measure to avoid abuse if the card gets stolen or lost. An ordinary ID card can only be used for identification, while a smart-card-based ID card can also be used to digitally sign documents and transactions in a nonrepudiated way (Rundgren, 1999a).

## Palmpad Computer and Laptop Computers

Palmpad and laptop computers have become another means of doing business over the Internet. Primarily developed to be portable and used for computing on the go, they are now being used for communication and access of information from databases at other locations. They originally could connect to the Internet via a mobile phone and conduct business. While laptops are fairly expensive, palmpads are less expensive in price and are becoming more common in m-commerce.

## Software

There are four basic components that make up a wireless Web service: browser phones, WML, link server, and services. Browser phones are handheld devices with special software that replaces conventional Web browsers. The WML is a programming language consisting of a set of statements that defines what the browser phone displays in its window and how it interacts with the user. Instead of Web pages, the wireless world uses decks consisting of cards (Vujosesevic & Laberge, 2000).

In 1995, European telecommunication companies wanted a common platform and decided on Java, which has today become the development language of choice for advanced cellular mobile phone services under the GSM digital communication platform. The advent of Java for the smart card computing environment, standardized as JavaCard application programming interface (API), now offers the prospect of an open mobile platform: one that can store multiple applications, as well as delete, replace, and upgrade them over the air, at the point-of-sale, or via the Internet. This technology gives operators new freedom to forge links with content providers, as well as develop unique applications and services (Brokat, 2000).

Wireless application protocol (WAP) is a framework specified by industry leaders in supporting mobile IT solutions. The WAP framework will be useful in digital cell phones in two ways: as a low-level communication protocol, and as an application environment supporting a "mini-browser." The latter will be used for sophisticated security-enabled applications like Internet banking and similar mission-critical information and transaction systems (Rundgren, 1999d).

Bluetooth wireless technology is a worldwide specification for a small-form factor, low-cost radio solution that provides links between mobile computers, mobile phones, other portable handheld devices, and connectivity to the Internet. It is the emerging short-range connection media of choice among mobile devices and electronic equipment. It works because it was developed as a cross-industry solution that marries a vision of engineering innovation with an understanding of business and consumer expectations. Along with a synchronization software, SyncML (an extensible and transport-independent technology), that allows a device to support a single synchronization standard for local synchronization over Infrared, Bluetooth, and USB as well as remote synchronization over Internet, WAP, and i-mode®, links can be established between groups of products simultaneously or between individual products and the Internet. This flexibility, combined with strict interoperability requirements, led to support for Bluetooth wireless technology from a wide range of market segments, including software developers, silicon vendors, peripheral and camera manufacturers, mobile PC manufacturers and handheld device developers, consumer electronics manufacturers, car manufacturers, and test and measurement equipment manufacturers (www.bluetooth.com).

# M-COMMERCE: CHALLENGES AND OPPORTUNITIES

M-commerce faces challenges, such as lack of standards, lack of ubiquitous wireless network coverage, technical differences among wireless devices, and security among others. Furthermore, the high prices of mobile services together with the slow access speed have not helped to add to the luster of the mobile environment. Some of the key challenges are as discussed below.

## Wireless Constraints

Developing content for wireless devices requires rethinking the Web experience. Wireless content developers need to begin from the ground up, developing content for these new devices. These devices tend to have little real estate available for viewing content—often as small as 14×7 characters. Wireless devices also tend to be monochromatic, so images do not render well. Keyboards are difficult to use. Wireless devices tend to have limited CPU, memory, and battery life. Developers and designers need to find new, intuitive navigational techniques to overcome these constraints. Today, the most common navigational technique on wireless is the drill-down capability (Gutzman, 2000).

Another constraint of wireless capabilities is the amount of bandwidth available for use of data transmission. This new technology would put a greater burden on current bandwidths available for wireless transmissions. Alternate bandwidths must be opened for transmission.

## Wireless User Behavior

Wireless users will not be expected to "surf the Web" in the traditional sense. This is due to the viewing and input constraints of using a wireless device and the relative inconvenience of performing any but the most straightforward, time-critical tasks. More likely, wireless users are expected to use their devices to execute small, specific tasks that they can take care of quickly, such as finding the time of local events, purchasing tickets, looking up news, or checking e-mail. Content developers need to develop with these motives in mind. Rather than just translating a content-rich site into WML, developers need to think in terms of surgical access to content and drilling-down capabilities to detailed information in the site (Gutzman, 2000).

Larger screens were developed for viewing, but use of magnification or projection techniques would make it easier for users to view Internet content. Keypads designed for smaller appliances should be developed with small typing ability in mind.

## Infrastructure for Wireless Internet

Currently, the infrastructure to handle smart cards is not generally established (except in Europe). Most industry analysts believe that smart cards will eventually become mainstream for paying in shops and on the Internet, together with a PC. In many countries, smart ID cards will also become fairly widespread. One of the problems is that the cost for shops, banks, companies, homes, and PC owners to convert to smart cards makes the process fairly slow. There is an obvious risk that consumers, banks, and companies, after the initial WAP-euphoria is gone, may start to question the rationale behind having multiple payment systems and could begin to put pressure on the mobile-phone makers to force them to adapt their systems to the rest of the world. This is an awkward solution, because it sets unnecessary physical constraints for mobile phones and is also likely to need "software fixes" for each new card variant. Even when used over GSM, operators will simply be supplying a gateway to the Internet, which will be regarded as a standard part of a subscription. Without such support, the m-commerce market could become severely crippled (Rundgren, 1999c).

## Security

Security of data transmissions and commerce being conducted by wireless devices is a great concern for businesses and individuals today. The wired Internet is vulnerable to interception or compromise by unauthorized users' attacks, because wireless LAN/WAN networks use a publicly available spectrum. Individuals have been wary of using Internet commerce for fear of having their credit card used improperly. A prerequisite for the success of m-commerce applications is the legal recognition and nondisputability of any transactions effected. The mobile digital signature may be an answer to this problem (Brokat, 2000).

New smart cards, available for wireless communications applications, will enable secure transactions via the Internet. The wireless identity module (WIM) will guarantee a new level of security by giving mobile Internet users the ability to safeguard their transactions through encryption and digital signatures. Compliant with the WAP, the WIM device will allow mobile network operators and service and content providers to begin implementing m-commerce services, such as secure information access, online banking, and the purchase of goods and services. The WAP-powered identity module supports "logical channels," enabling users to pass from one application to another without losing transactions that have already been carried out. The card offers two forms of protection: client-to-server authentication using ultralong keys, and the ability to generate the digital signature required to secure the application. Unlike an encryption-enabled browser, the secret keys handling the encryption remain in the user's smart card, by definition a tamper-resistant device, and allow it to be removed and transferred to other devices (*Electronic Buyer's News*, 1999).

One advantage of smart card IDs is that they are extremely hard to forge. To crack a private key stored in a smart card or guess its value based on a corresponding public key is very difficult. A PIN code is added as an extra security measure to avoid abuse if the card gets stolen or lost. An ordinary ID card can only be used for identification, while a smart-card-based ID card can also be used to digitally sign documents and transactions in a nonrepudiated way (Rundgren, 1999a). However, the security measures typically implemented with a wireless application delivery approach can add to the costs and make the computing system more complex to administer and use.

## Privacy

Privacy is another issue not resolved by the growth of m-commerce. The new connectivity of consumers to the Internet is a great convenience for consumers, but it also comes at a price. The price is the value of privacy that individuals lose, as they become hooked-up to the Internet. One part of privacy is that the development of smart cards for use with cell phones is convenient for consumers wanting to buy or sell. However, much personal data is enclosed on the card, and it could be used for the wrong purposes. Many cell phones can be equipped with a global positioning chip that can identify the location of the user. This new technology would be good for emergencies but could also be used against the individual for monitoring purposes or other activities. These are issues that still need to be addressed and have been downplayed by current technology developers. Privacy is one of several issues that complicate the long-term timetable for developing location-based m-commerce. Another issue is the level of direct access marketers will have to customers, because the Internet will be located with the individual customer and can be contacted by voice, e-mail, or Internet (Vujovesic & Laberge, 2000). Next, we discuss the opportunities made available by m-commerce and wireless communications.

## Opportunities

Despite the barriers that still exist for the unfettered deployment of m-commerce, cautious optimism for growth is pervasive. One of the driving forces for increased use of mobile devices is the desire to change business processes inside the enterprise, rather than between businesses or based solely on a consumer model as prevailing predictions from several years ago.

A current survey of businesses by Forrester Research (Cambridge, Massachusetts, USA) found that 47% of the country's leading companies are in the process of rolling out mobile applications or are considering doing so (Mertz & Serrell, 2002). PricewaterhouseCoopers, in their 2002-2004 Tech Forecast, projected worldwide revenue for mobile middleware to grow 60.7% compounded annually between 2000 and 2005 (PricewaterhouseCoopers, 2002). Applications for remote/off-site employees (viz., salespeople, truckers, and so forth) also provide a significant opportunity. A popular innovation in the United States is the satellite tracking and communication devices in approximately 1 million vehicles, such as the OnStar® equipment from GM®. OnStar has 800,000 subscribers in the United States and Canada and expects 4 million vehicles to be equipped with this system by 2003 (Kalakota & Robinson, 2001, p. 6).

With the packet-switching technology in the GPRS network, carriers can optimize how they transmit data, sending data when dips in the capacity utilization of the networks occur. By sending data at a lower cost, carriers can stimulate demand and extract higher profit margins by selling wireless data services. Pricing, coverage, and functionality are the key drivers for consumers in the United States to adopt new wireless technologies and services. To convince customers to adopt wireless, carriers will have to embrace pricing models, such as the pay-per-use billing model made possible by packet-switched networks. Accenture Inc. (New York, USA), an information technology consulting and management company, in its recent report titled "The Future of Wireless," predicts that the global market for small wireless Internet-capable devices, such as handheld computer, basic microbrowser phones, and smart phones will grow 630% by 2005 (Beal et al., 2001). This projection was based on the current level of mobile devices and the mobile commerce technology available in the market. Thus, this projection could increase dramatically with the arrival of new wireless technology.

While most of the industry foresight focused on mobile solutions for the enterprise arena, PricewaterhouseCoopers (now IBM) weighed in projecting both collaborative use (between enterprises) as well as consumer use. The upshot is that while enterprise solutions will maintain the lion's share of the m-business market, with the fastest growth and majority of revenue, collaborative and consumer markets will also grow exponentially over the next few years. Listed in Appendix B is the growth potential forecasted across the world by various research organizations.

# PREDICTIONS OF FUTURE GROWTH OF M-COMMERCE

In the next three years, we will see the convergence of two major trends: "software as a service" and the rapid growth of wireless network and device capabilities. Software as a service is the provision of access to integrated, centralized applications and data to any location at any time and is a trend that has grown out of the need of an increasingly mobile enterprise workforce. The wireless LANs and WANs will help deploy a wide range of applications, such as customer relationship management (CRM) and groupware to meet business-to-employee, business-to-business, and business-to-consumer needs (Citrix Systems, Inc., 2001).

The value of transactions conducted over mobile phones in Europe is set to reach 23 billion euros ($23.7 billion) by 2003, according to a new study from Durlacher Research Ltd. (London). Europe's mobile phone service operators are poised to increasingly derive revenue from Internet content and services, and will become leading Internet portals in the future. Europe adopted a clear lead in usage and application development, fueled by its high penetration of mobile phones and successful adoption of GSM as the single digital phone standard. The United States has not been able to reach a single standard or settle on a generic type of terminal, thus slowing the establishment of a critical mass of handsets in the market needed for the introduction of new services (Uimonen, 1999).

According to a recent study by IDC, the number of people using wireless devices to connect to the Internet will increase by some 728% by 2003. That is an increase from 7.4 million users in 1999 to 61.5 million users in 2003 (Blackwell, 2000).

Mobile wireless devices are a reality today, with close to a billion subscribers worldwide and expectations of up to 1.5 billion users by the year 2005 (Van Impe, 2002). Illustrated in Appendix A, Table 1, is the penetration of mobile data users by region, and in Appendix A, Figure 1, the worldwide Internet access by device is shown. In 2003, around a quarter of all mobile Internet users are likely to use their mobile phone to access travel services such as booking flights, finding local hotel accommodations, sourcing last-minute holidays, or purchasing rail tickets. These are all services that are particularly suited to business travelers and experienced leisure travelers. This mix of mobile transactions is likely to result in mobile travel commerce revenues overtaking those of online travel (Cross, 2000).

Smart cards have always offered the potential to radically change and automate the way mass consumer business operates. The huge business potential presented by Java-based smart cards to service providers has not

*Figure 3: Worldwide Mobile Commerce Use*

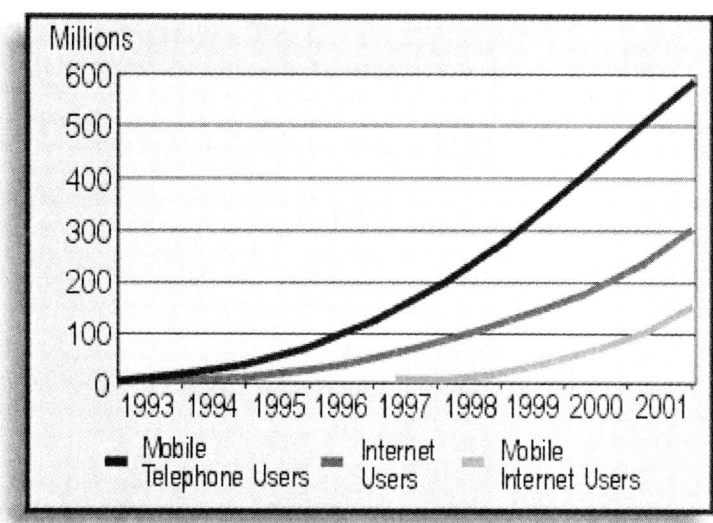

*Figure 4: Worldwide Handset Sales*

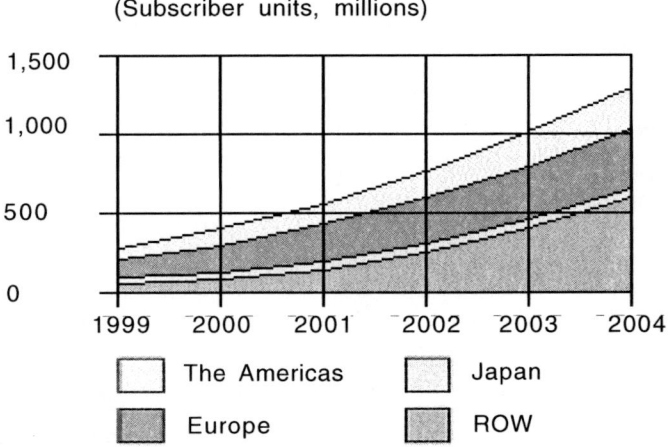

(Subscriber units, millions)

Legend: The Americas, Europe, Japan, ROW

*Source: Cahners In-Stat Group*

been lost to GSM-based mobile communications operators. Today, the cost of mobile communications is almost the same as fixed-line communications. Mobile phone penetration in Asia is expected to reach 35% to 40%. With the merging of electronic tools, such as the palmtop with the mobile phone, and more use of the mobile phone for accessing data, the mobile phone complements the Internet to enable Web surfers to access information without needing a PC. Eventually, e-commerce will be a hot application for mobile phone users. The same should be expected for GSM-based mobile e-commerce. The capability of mobile phones is expected to increase, and this will accelerate more developments (Brokat, 2000).

# IMPLICATIONS FOR MANAGEMENT

Consumer-oriented m-commerce is becoming a reality today. Many businesses and consumers are taking up the wireless Web through many services, such as AT&T, Sprint PCS, Verizon Wireless, Motorola, Nokia, Ericsson, and other wireless service providers. Personalizing, localizing, and customizing the content and services to the individual needs of the mobile device users is the key to success of its commercialization. This focus on relevancy (i.e., showing customers what they can do to enhance productivity and add convenience in their daily lives) represents a theoretical change for equipment makers and carriers and proves that a more conservative marketplace can embrace technology en masse only when it is relevant for the majority of unique users.

It will still take a few years for consumer-oriented m-commerce applications to become as generally available as the wired Web is today. In the near term, the most promising opportunities for mobile wireless transactions are those built for industrial use. These involve the development of software for vertical applications that allow delivery agents, salespeople, and mobile workers to perform logistical and other data-driven duties. Doctors are using wireless-enabled Palms to access and update patient records or write prescriptions. The most visible wireless developments are consumer oriented. These include delivery of time-critical information to mobile banking and travel-ticket purchase. The current crop of consumer-oriented services will become the foundation for more advanced services that will deliver time- and location-critical data to consumers (Vujovesic & Laberge, 2000).

M-commerce has features inherent to its nature, such as ubiquity, flexibility, convenience, and personalization, and today's features of mobile communication open a world of opportunities for m-commerce tomorrow. However, only those companies with the ability to use these inherent qualities of m-commerce to offer value to their customers will succeed in the mobile environment. The mobile environment still has the potential to empower people, providing them with real-time wireless applications that will make their lives easier and business more efficient and productive.

M-commerce is a reality now and will not be going away in the near term. Problems with these systems are being addressed, and new applications are rapidly being developed. Most of the market is behind this new technology, and it will likely change business by making it easier and more accessible to individuals. M-commerce is a tool of telecommunication and Internet industries. Those involved in it now, may become the giants that Microsoft and Intel have been in the computing industry. Those who follow may be able to capitalize on leading m-commerce mistakes and perfect this technology. One should measure the risk involved and understand that a carefully planned strategy is necessary to implement this new technology into corporate future goals.

# CONCLUSION

The development of m-commerce and the wireless World Wide Web is the evolution of several different technologies to make the Internet more accessible and commerce easier for the consumer. While the Internet is already a valuable form of business, which has already changed the way the world is doing business, the format in which we will view it is changing. There are many new opportunities that have only just begun to be explored. This will become a large opportunity for those who capitalize upon this technology. The growth trends are impressive, and the public interest and large companies are behind this technology.

We are currently in a wireless version of the chicken-and-the-egg scenario. Finding a stable, high-bandwidth wireless option is critical if wireless technology is to become a fundamental part of our day-to-day lives. However, the carriers need to know if the technology will be accepted before investing billions in new infrastructure. Acceptance will be determined by how the technology can be applied to the users' day-to-day activities. New networks of wireless networks are springing up around the country, as users find a need for wireless applications that are not currently supported by the existing wireless companies.

The application of m-commerce technology is the true seller. Its success is contingent upon a majority of Internet browsers using mobile digital phones. To be fully accepted, all of these technologies must overcome their current drawbacks. Technology is being developed to overcome the security drawbacks, but enhanced viewing devices and input devices for controlling the data must be developed. As we go toward the new generation of wireless telecommunications, the United States, in order to keep it from lagging behind Japan and Europe, needs a single technical standard for newer generations of wireless devices and a coherent electromagnetic spectrum allocation policy that would allow for more and cheaper spectrum to be used for telecommunications. Also, the infrastructure to control smart card payments may be a few years off for the United States, but it will need to be accepted at shops and businesses throughout the United States to make it useful. Mobile e-commerce will change our world of business to a similar degree that the Internet alone has changed business today.

# REFERENCES

Beal, A., Beck, J. C., Keating, S. T., Lynch, P. D., Tu, L., Wade, M., & Wilson, J. (2001). *The future of wireless: Different than you think, bolder than you imagine.* Accenture, Institute for Strategic Change, June 4. Available at: http://www.accenture.com/xd/xd.asp?it=enWeb &xd=_isc/iscresearchreportabstract_134.xml.

Blackwell, G. (2000). *Wireless to outstrip wired net access.* Available at: http:/www.isp-planet.com/research/more_wireless.html.

Citrix Systems, Inc. (2001). Wireless application access white paper.

Cross, T. (2000). *Mobile travel commerce—A bigger deal than online travel?* Available at: http:/www.gmcforum.com/PressRelease/ PressRelease_110500.htm.

Duffey, K. (1997). *Mobile commerce* [Online]. Available at: http:/ public.logica.com/~mcommerce/ourvisio.htm.

Duffey, K. (1998). *The new wireless age* [Online]. Available at: http:/ www.singapore.cnet.com/Ebusiness/Ecommerce/Mcommerce/ss01.html.

Durlacher Research. (1998). *Mobile electronic commerce* [Online]. Available at: http:/network365.com/mobilecommerce.html.

*Electronic Buyer's News.* (1999). *Schlumberger says new smart card will ensure secure mobile-Internet transactions* [Online]. Available at: http: /www.ebns.com/ecomponents/commnews/story/OEG19991116S0008.

*Eurotechnology-Japan.* (2000). *How many Internet users are there in Japan?* [Online]. Available at: http:/www.Eurotechnology.com.

Gutzman, A. (2000). *The who, what and why of WAP* [Online]. Available at: http://www.allnetdevices.com/wireless/opinions/2000/06/20/ the_who.html.

Hansen, C. (2000). *GSM-based mobile e-commerce will be hot* [Online]. Available at: http:/www.globalsources.com/MAGAZINE/TS/9909/ SLB.HTM.

Intel. (1999). The future GSM data knowledge [Online]. Available at: http:/ www.gsmdata.com/Future.html.

Kalakota, R. & Robinson, M. (2001). *Mbusiness: The Race to Mobility* (p. 6). New York: McGraw-Hill.

Mertz, C., & Serrell, M.D. (2002). Mobile application tools. *PC Magazine Online,* (October 15). Available at: http://www.pcmag.com/article2/ 0,4149,545121,00.asp.

*Mobile Business.* (2000). *Brokat global e-commerce services* [Online]. Available at: http:/www.brokat.com/int/mobile/index.html.

Muller, J. & Schnoring, T. (1995). *Mobile Telecommunications: Emerging European Markets* (p. 247). Norwood, MA: Artech House.

PriceWaterhouseCoopers. (2002). *Technology forecast: 2002-2004*, p. 680.

Reiter, A. (1999a). *Dynamics of wireless e-commerce, conditions sparking the development of international wireless e-commerce* [Online]. Available at: http:/www.wirelessinternet.com/dynamics.htm.

Reiter, A. (1999b). *Wireless e-commerce: A new business model* [Online]. Available at: http:/www.wirelessinternet.com/wireless2.htm.

Rundgren, A. (1999a). *ID-cards: Yesterday, today and in the future* [Online]. Available at: http:/www.mobilephones-tng.com/papers/idcards.html.

Rundgren, A. (1999b). *The cyber ID card* [Online]. Available at: http:/www.mobilephones-tng.com/v100/cyberphonecards.html.

Rundgren, A. (1999c). *The new Swiss Army Knife? (Smart cards vs. smart terminals)* [Online]. Available at: http./www.mobilephones-tng.com/papers/thenewswissarmyknife.htm.

Rundgren, A. (1999d). *WAP—Wireless Application Protocol* [Online]. Available at: http:/www.mobilephones-tng.com/v100/wap.htm.

Uimonen, T. (1999). *European mobile commerce to hit $24 billion* [Online]. Available at: http:/www.durlacher.com.

Van Impe, M. (2002). Nokia expects the number of mobile users to surge in the next three years. *Mobile CommerceNet*, (June 19). Available at: http://www.mobile.commerce.net/story.php?story_id=1824.

Varshney, U. & Vetter, R. (2000). Emerging mobile & wireless networks. *Communications of the ACM, 43*(6).

Vujosevic, S. & Laberge, R. (2000). *Info on the go: Wireless Internet database connectivity with ASP, XML, and SQL server* [Online]. Available at: http:/www.msdn.microsoft.com/msdnmag/issues/0600/wireless/wireless.asp.

Walker, M. (2000). *M-commerce tricks emerge from tech magician's bag* [Online]. Available at: http:/www.bizjournals.com/houston/stories/2000/06/26/focus6.html.

# APPENDIX A

*Table 1: Penetration of Mobile Data Users by Region*

|  | 2000 | 2003 | 2005 |
|---|---|---|---|
| United States | 7% | 44% | 83% |
| Japan | 21% | 62% | 90% |
| Asia-Pacific | 1% | 4% | 8% |
| Western Europe | 23% | 72% | 91% |
| Rest of World | 0% | 4% | 10% |
| Total | 3% | 12% | 20% |

*Source: ARC Group, 2000*

*Figure 1: Worldwide Internet Access by Device (1999-2004)*

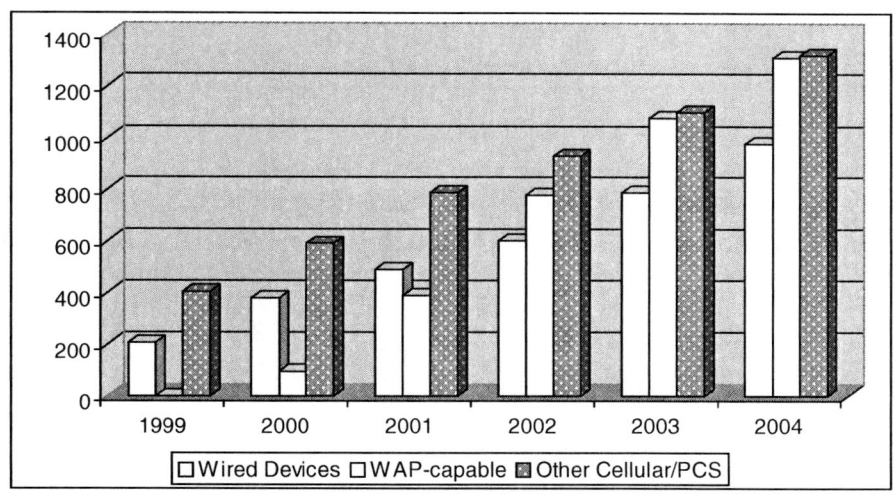

*Source: International Data Corporation, 2001*

# APPENDIX B:
# KEY M-COMMERCE STATISTICS

**Services Used By Internet-Enabled Mobile Phone Users Globally**

| Country (as % of IE-mobile users) | E-mail | Banking | Purchasing | Games |
|---|---|---|---|---|
| Asia | 10% | 2% | 3% | 3% |
| Brazil | 11% | 7% | 1% | 2% |
| Europe | 10% | 3% | 1% | 3% |
| Japan | 77% | 4% | 12% | 5% |
| North America | 27% | 6% | 3% | 7% |
| Worldwide | 19% | 3% | 3% | 3% |

Source: AT Kearney, August 2002

**Predicted m-Commerce Revenues, 2001-2005**

| Research Entity (USD billion) | 2001 | 2002 | 2003 | 2004 | 2005 |
|---|---|---|---|---|---|
| Datamonitor (2000) | 1.5 | 3.5 | 5.0 | 7.5 | 8.5 |
| Forrester Research (2000) | 1.0 | 2.5 | 7.5 | 14.0 | 22.0 |
| Durlacher (realistic/interpreted) | 3.0 | 3.5 | 5.1 | 10.0 | 19.0 |
| Frost & Sullivan (1999) | 8.0 | 10.0 | 15.0 | 19.0 | 24.0 |
| Consult Hyperion (seven countries, 2001) | 2.5 | 3.5 | 4.5 | 5.5 | 7.5 |
| Jupiter Research (2000) | 1.0 | 2.0 | 3.0 | 5.0 | 8.0 |

Source: Transaction Magazine, August 2002

**Prepaid Subscribers In World Regions, 2001-2004**

| Region | 2001 | 2002 | 2003 | 2004 |
|---|---|---|---|---|
| Europe | 258,1 | 320,9 | 375,6 | 427,6 |
| Asia Pacific | 27,7 | 40,7 | 55,4 | 71,3 |
| Greater China | 55,4 | 93,4 | 119,2 | 156,8 |
| North America | 12,5 | 15,7 | 18,4 | 20,3 |
| South America | 63,2 | 86,1 | 109,7 | 133 |
| Worldwide | 416,2 | 556,8 | 678,3 | 809 |

Source: EMC

**Current Mobile Phone Users' Interest In 3G Applications**

| Application | W Europe | E Europe | USA |
|---|---|---|---|
| On 6-point interest scale, 6 = high interest, and 1 = low interest | | | |
| E-mail | 4.5 | 4.7 | 4.3 |
| Payment Authorization/ Enablement | 3.4 | 3.8 | 3.0 |
| Banking/Trading Online | 3.5 | 3.4 | 3.2 |
| Shopping/Reservations | 3.0 | 3.1 | 2.9 |
| Interactive Games | 2.0 | 2.2 | 2.4 |

Source: Taylor Nelson Sofres, May 2002

**Global Internet and Wireless Users, 2001, 2004 and 2007**

| Subscribers | 2001 | 2004 | 2007 |
|---|---|---|---|
| Internet users (millions) | 533 | 945 | 1,460 |
| Wireless Internet users as % of all Internet users | 16 | 41.5 | 56.8 |

Source: eMarketer, March 2002

**Proportion Of Wireless Web Users, By Country, 2001**

| Country | Percentage |
|---|---|
| United States | 6 |
| United Kingdom | 10 |
| Germany | 16 |
| Finland | 6 |
| Japan | 72 |

Source: Accenture (summer 2001)

**m-Commerce Comfort Levels of US Consumers (per device)**

| Status | Phones | PDAs |
|---|---|---|
| Not at all comfortable | 65% | 61% |
| Uncomfortable | 21% | 24% |
| Neutral | 6% | 7% |
| Comfortable | 6% | 6% |
| Extremely comfortable | 2% | 2% |

Source: Forrester Research

| Worldwide Number of Mobile Users And Consumers, 2001 | | |
|---|---|---|
| Region | Mobile Users | Mobile Consumers |
| Africa | 4,900,001 | 1,650,000 |
| Asia-Pacific | 206,500,000 | 131,750,000 |
| Europe | 68,850,000 | 39,350,000 |
| Central/South America | 18,250,000 | 11,850,000 |
| North America | 133,290,000 | 86,790,000 |
| Australia | 5,250,000 | 3,100,000 |
| Former USSR | 11,191,500 | 8,191,500 |
| World | 448,231,500 | 282,681,500 |
| Source: ResearchPortal.com, via CyberAtlas | | |

| US Mobile Commerce Revenues, 2001-2007 | |
|---|---|
| Year | Total (USD) |
| 2001 | 127 million |
| 2002 | 616 million |
| 2003 | 2.1 billion |
| 2004 | 5.7 billion |
| 2005 | 13.1 billion |
| 2006 | 29.0 billion |
| 2007 | 58.4 billion |
| Sources: IDC and Jupiter Media Metrix | |

**Obstacles Preventing Consumers From Adopting m-Commerce**

| Obstacle | Phones | PDAs |
|---|---|---|
| Credit card security concerns | 52% | 47% |
| Fear of 'klunky' user experience | 35% | 31% |
| Don't understand how it would work | 16% | 16% |
| Other | 11% | 13% |
| Never heard of it before | 10% | 12% |

Source: Forrester Research

**Worldwide Users of Wireless Financial Payments, 1999-2004**

| (in millions) | 1999 | 2000 | 2001 | 2002 | 2003 | 2004 |
|---|---|---|---|---|---|---|
| Europe | 0.5 | 2 | 4 | 10 | 19 | 31 |
| Asia-Pacific | 0.5 | 3 | 7 | 13 | 21 | 29 |
| US | n/a | n/a | n/a | 0.5 | 1 | 2 |

Source: Celent Communications

**Users of Wireless Financial Services In World Regions, 2000-2005**

| Region | 2000 | 2005 |
|---|---|---|
| North America | 0.45 mn | 34.97 mn |
| Western Europe | 3.89 mn | 76.55 mn |
| Asia-Pacific | 4.81 mn | 83.74 mn |

Source: TowerGroup

**Percentage of WAP-Enabled Banking in Western Europe, 2000**

| Country | Total (per cent) |
|---|---|
| Scandinavia | 48% |
| UK | 22% |
| Germany | 13% |
| France | 5% |
| Italy, Spain | 5% |
| Others | 7% |

Source: International Data Corporation

**WAP-Enabled Bank Accounts in Western Europe, 2000-2004**

| Year | Total (millions) |
|---|---|
| 2000 | 2 |
| 2001 | 5 |
| 2002 | 13 |
| 2003 | 24 |
| 2004 | 32 |

Source: International Data Corporation

**North American & Global Wireless Web Users, 2000-2005**

| Region | 2000 | 2001 | 2002 | 2003 | 2004 | 2005 |
|---|---|---|---|---|---|---|
| North America | 0.2 mn | 0.3 mn | 14 mn | 37.5 mn | 63.7 mn | 95.6 mn |
| World Total | 6 mn | 16 mn | 77 mn | 190 mn | 322 mn | 484 mn |

Source: Ovum

**Global Mobile Commerce Revenues, 2000 - 2005 (USD millions)**

| Region | 2000 | 2001 | 2002 | 2003 | 2004 | 2005 |
|---|---|---|---|---|---|---|
| North America | 0.0 | 0.1 | 0.2 | 0.7 | 1.8 | 3.5 |
| Western Europe | 0.0 | 0.1 | 0.5 | 1.7 | 4.6 | 7.8 |
| Asia | 0.4 | 1.3 | 2.6 | 5.0 | 7.4 | 9.4 |
| Latin America | 0.0 | 0.0 | 0.0 | 0.1 | 0.2 | 0.5 |
| Other | 0.0 | 0.0 | 0.1 | 0.2 | 0.4 | 1.0 |
| **Global** | **0.4** | **1.5** | **3.4** | **7.6** | **14.5** | **22.2** |
| US | 0.0 | 0.1 | 0.2 | 0.6 | 1.7 | 3.3 |
| Japan | 0.4 | 1.2 | 2.1 | 3.5 | 4.5 | 5.5 |

Source: Jupiter Research

**Percentage of Wireless Web Users By Region, 1998-2002**

| Region (millions) | 1998 | 1999 | 2000 | 2001 | 2002 |
|---|---|---|---|---|---|
| Europe | - | 7 | 20 | 62 | 130 |
| US | 9 | 19 | 40 | 62 | 80 |
| Japan | 9 | 17 | 20 | 30 | 50 |

Source: *Forbes Magazine*

**Global Users of 3G Wireless Data in 2011**

| Country | Total |
|---|---|
| Europe | 251 million |
| China | 247 million |
| United States | 208 million |
| Japan | 96 million |

Source: Kalba International

**Numbers Using Wireless Data in the US By 2005**

| Category | Total (millions) |
|---|---|
| Entire user base | 9.5 |
| B2B subscribers | 6.8 |
| Residential subscribers | 2.8 |
| Global Web-enabled devices | 500 |

Source: The Strategis Group

**Smart Phones Versus Conventional Wireless Phones (in millions)**

| | 1999 | 2000 | 2001 | 2002 | 2003 |
|---|---|---|---|---|---|
| Smart Phones | 0.2 | 6 | 56 | 175 | 330 |
| Conventional | 250 | 304 | 284 | 185 | 40 |

Source: Datacomm Research

**Uptake of Wireless Data Applications, 2000-2005 (in millions)**

| Applications | 2000 | 2001 | 2002 | 2003 | 2004 | 2005 |
|---|---|---|---|---|---|---|
| Messaging | 100 | 230 | 399 | 611 | 916 | 1268 |
| E-commerce/Retail | 12 | 36 | 107 | 195 | 318 | 469 |
| Financial Services | 50 | 123 | 225 | 357 | 529 | 798 |
| Intranet (corporate) | 5 | 20 | 49 | 81 | 129 | 206 |
| Internet, WAP | 4 | 20 | 85 | 183 | 344 | 614 |
| Entertainment | 61 | 143 | 246 | 372 | 554 | 775 |
| Navigation | 47 | 146 | 239 | 345 | 488 | 785 |

Source: allNetDevices

**Global Internet-Enabled Mobile Terminals, 1999-2003 (millions)**

| Year | Europe | Asia-Pacific | US/Canada | Latin America | Middle East | Africa |
|---|---|---|---|---|---|---|
| 1999 | 0.26 | 4.0 | < 0.04 | 0.0 | < 0.01 | 0.0 |
| 2000 | 20.24 | 13.55 | 6.50 | 1.27 | 2.00 | 0.90 |
| 2001 | 104.58 | 48.72 | 42.23 | 12.04 | 10.90 | 4.86 |
| 2002 | 204.66 | 119.01 | 103.33 | 44.40 | 30.13 | 13.56 |
| 2003 | 273.45 | 187.19 | 158.69 | 101.48 | 54.04 | 23.68 |

Source: Dataquest

**Total Number of WAP-Enabled Subscriptions in Europe (in millions)**

| Region | 1999 | 2000 | 2001 | 2002 | 2003 | 2004 | 2005 |
|---|---|---|---|---|---|---|---|
| Northern Europe | n/a | 2 | 3 | 8 | 10 | 12 | 15 |
| Southern Europe | n/a | 3 | 12 | 21 | 30 | 40 | 50 |
| Central W Europe | n/a | 10 | 30 | 58 | 82 | 100 | 122 |

Source: Datamonitor

**Total Number of Cellular Subscriptions in Europe (in millions)**

| Region | 1999 | 2000 | 2001 | 2002 | 2003 | 2004 | 2005 |
|---|---|---|---|---|---|---|---|
| Northern Europe | 15 | 20 | 24 | 24 | 25 | 25 | 25 |
| Southern Europe | 48 | 58 | 68 | 70 | 75 | 78 | 80 |
| Central W Europe | 70 | 100 | 140 | 160 | 170 | 175 | 178 |

Source: Datamonitor

**Worldwide Penetration of Handheld Devices Through 2005**

| Country | 1999 | 2005 |
|---|---|---|
| Asia | 125 million | 310 million |
| Germany | 22 million | 62 million |
| UK | 21 million | 45 million |
| France | 17 million | 45 million |
| Netherlands | 7 million | 12 million |
| Belgium | 3 million | 7 million |
| Austria | 3 million | 6 million |
| US | 1.7 million | 24 million |

Source: Datamonitor

**Chapter IV**

# Opportunities and Limitations in M-Commerce

P.W. Lei, University of Sussex, United Kingdom

C.R. Chatwin, University of Sussex, United Kingdom

R.C.D. Young, University of Sussex, United Kingdom

S.H. Tóng, Instituto Superior de Ciências do Trabalho
e da Empresa, Portugal

## ABSTRACT

*Electronic commerce (e-commerce) activity is growing exponentially, and it is revolutionizing the way that businesses are run. There is now an explosion of mobile wireless services accessible via mobile phones and Personal Digital Assistants (PDAs). Mobile e-commerce (m-commerce) makes business mobility a reality. Mobile users can access the Internet at any time, from anywhere (even from their shirt pockets/purses) using ubiquitous inexpensive computing. It is estimated that the m-commerce* *market was worth US$3.5 billion in 2000 and will grow to over US$200 billion by 2005 (Abbott, 2002). M-commerce is generally considered to be an extension of e-commerce. In fact, m-commerce has unique characteristics and functionality. Hence, it creates a unique and new business opportunity. Tesco, the United Kingdom-based supermarket, rolled out their mobile service, but the U.S. bank, Wells Fargo, is planning to close down their mobile service later this year due to lack of interest. M-commerce has a number of inherent complexities, as it embraces many emerging technologies: mobile wireless systems, mobile handheld devices, software, wireless protocols, and security. These technologies have rapid product cycles and quick obsolescence. In this chapter, we will examine the opportunities and limitations of m-commerce and concentrate our discussion on mobile phone systems.*

# INTRODUCTION

The airwaves are crammed with various mobile wireless services, ranging from mobile phone networks to wireless local area networks (LANs). There are many definitions of mobile commerce (m-commerce). One definition consists of all or part of mobile wireless services—the service provided by mobile phone systems, which has achieved huge success. Mobile phone users originate from all walks of life and include almost all age groups—from teenagers to retired people. It creates a new way of personal communication without location constraints. The number of active mobile phones grew tenfold in the period from 1996-2001, and the growth is still strong, despite the economic downturn. The number of active cell phones was around 946 million in 2001 (Shirky, 2002). In addition, each wireless service has unique technology. Instead of briefly describing all mobile wireless services, we will concentrate on the mobile phone and the PDA related to mobile telecommunication. M-commerce is, therefore, defined as e-commerce carried out in handheld devices, such as the mobile phone and PDA, through a mobile telecommunication network. The chapter is organized as follows. The first two sections introduce the main technologies—the mobile telecommunication network and the wireless protocols, where the challenges for m-commerce are investigated. Then m-commerce applications are discussed. Finally, we will highlight the issues in implementing m-commerce.

# MOBILE TELECOMMUNICATION

## Mobile Network Generations

The first-generation (1G) cellular systems were the Advanced Mobile Phone Systems (AMPS) introduced in 1983 in the United States. The 1G systems are analog, circuit-based, narrowband, and are suitable for voice communication only. In the 1990s, the second-generation (2G) systems were introduced. The 2G systems are digital, circuit-based, and narrowband. The 2G system is primarily for voice communication with limited data communications. The major 2G networks are based on the technologies of Code Division Multiple Access (CDMA), Global System for Mobile communications (GSM), and Personal Digital Cellular (PDC).

Many mobile operators are currently upgrading the 2G to 2.5G or 3G. Some 2.5G systems were implemented using General Packet Radio Service (GPRS). GPRS is a packet-based extension of GSM and provides high data throughout. The objective of integrating GPRS into GSM is to increase the number of connections per bearer by utilizing the given physical channels more efficiently than the existing service. Packet switching means that GPRS radio resources are only used when users are actually sending or receiving data. Hence, the scarce radio resources are being utilized efficiently, and the data rate is 64-144 kbps. It brings Internet protocol (IP) capability to the GSM network. The GSM/GPRS systems will then evolve into Enhanced Data for GSM Evolution (EDGE) and finally into the Universal Mobile Telecommunication System (UMTS), the third generation (3G).

The 3G technology comprises three primary standards: W-CDMA (wideband CDMA), CDMA2000, and TD-CDMA (time division CDMA). Most of Europe and Japan settled on W-CDMA, an upgrade to the GSM standard, which is widely used in those areas (Garber, 2002). TD-CDMA, developed in China, will be used there only. The United States is working with all three major 3G standards. However, AT&T, the largest mobile operator in the United States, is upgrading their networks to W-CDMA, which will be the dominant global 3G technology. 3G's data rate is up to 2 mbps when stationary, less when moving. 3G networks provide higher-speed transmission to support high-quality audio and video, as well as a global roaming capability. This means that the same mobile phone can be used in London, New York, and Tokyo. Shown in Table 1 is a comparison of 2.5G with 3G in services and data rates. In Japan, NTT DoCoMo rolled out 3G in May 2001. Hutchison rolls out the first U.K. 3G system in March 2003. Verizon launched the first U.S. 3G network in January 2002. There are two devices available. One is a handset

*Table 1: A Comparsion Between 2.5G and 3G*

| 2.5G (GPRS) | 3G |
|---|---|
| • Web surfing up to 171 kbps | • Superfast Internet access up to 2 mbps |
| • E-mail and picture message | • Live audio and video broadcasting |

and has to be cabled to a computer. The other is a wireless modem in a PDA. A 3G phone is not available. In the mobile Internet or m-commerce, Japan and Europe lead the world, rather than the United States, the leader in Internet— e-commerce.

## GSM Mobile Communication Architecture

Cellular networks are based on the technique of frequency reuse, so that the limited radio spectrum will receive maximum use. Illustrated in Figure 1 is frequency reuse. In cellular radio networks, the area covered by one base station is reduced, and other base stations are installed with small overlapping areas. Adjacent cells need to use different frequencies to avoid interference, but the same frequency can be reused in nonadjacent cells. The coverage area is split into many cells. By splitting an area into many smaller cells, the overall

*Figure 1: Frequency Reuse*

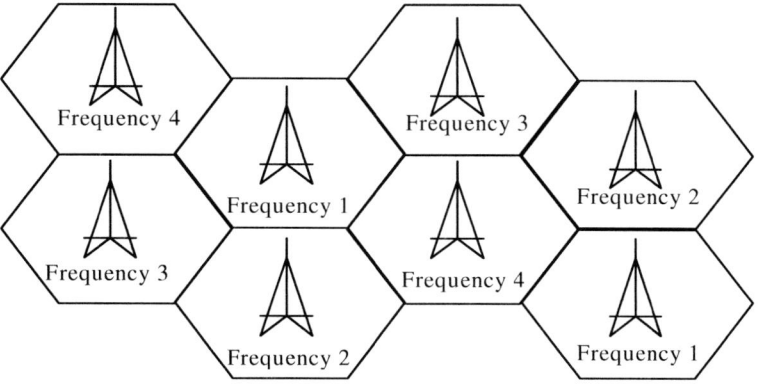

network capacity increases, but a decrease in the reuse of frequency is also seen. The major factors in determining the optimal size of each cell to be considered are as follows:

- User density.
- Area characteristics (urban, suburban, rural).
- Presence of buildings.

GSM is widely implemented in 178 countries and accounts for more than 70% of mobile phone users in the world (UMTS Forum, 2002). The GSM is composed of three main elements: the switching subsystem, the base station subsystem, and the mobile station. In Figure 2, an overview of the complete GSM system is presented. A mobile station is carried by the subscriber, such as a mobile phone. A GSM mobile phone consists of two parts. The first part contains the hardware and software relating to the mobile interface. The second part is the Subscriber Identity Module (SIM) card, used to store the subscriber's

*Figure 2: Overview of the Complete GSM System*

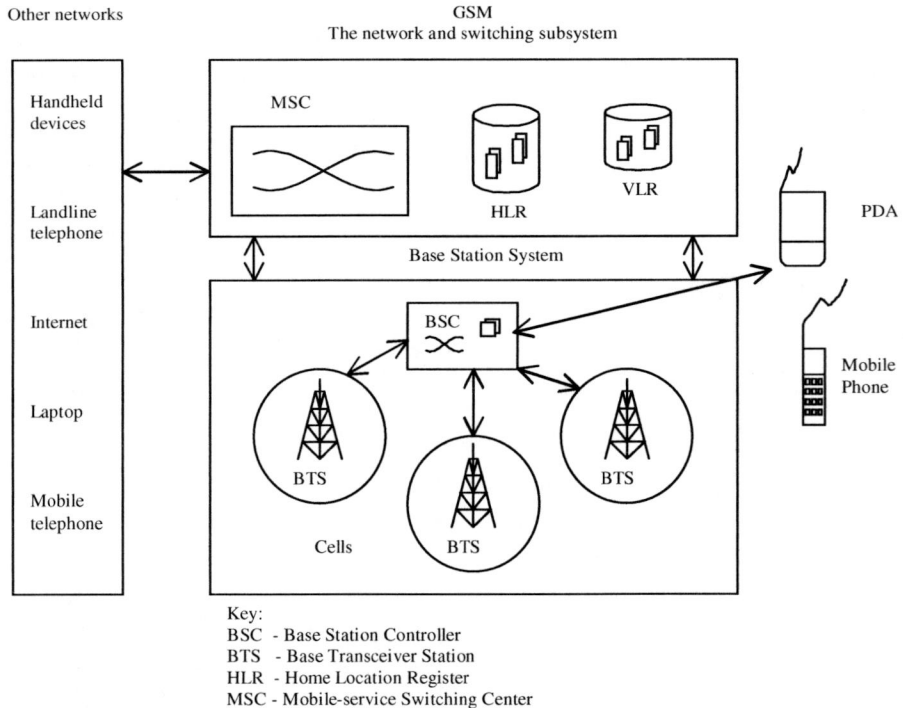

Key:
BSC  - Base Station Controller
BTS  - Base Transceiver Station
HLR  - Home Location Register
MSC - Mobile-service Switching Center
VLR  - Visitor Location Register

personal data. The subscriber uses the SIM card to check into the mobile network. It contains the secret keys needed by the GSM security algorithms for authentication and encryption. Also, the SIM card comes with a CPU, EEPROM memory for the file system, a ROM memory for the card operating system, and a RAM memory for program execution. Java applications can be run on the card itself.

The base station subsystem consists of BTS (Base Transceiver Station) and BSC (Base Station Controller). The BTS comprises the transmitting and receiving facilities, including antennas and all the signaling related to the radio interface. A BSC generally manages several BTSs and is linked to the switching subsystem, which forms the gateway network between the mobile network and the wired networks, such as the public switched telephony network. The switching subsystem consists of the mobile-service switching center (MSC) and a number of databases—the Home Location Register (HLR) and the Visitor Location Register (VLR) (Walke, 2002). It provides all the functionality needed to handle a mobile subscriber. The HLR is concerned with the billing of subscribers, and the VLR is used to manage the subscribers currently roaming in the area under the control of the MSC.

The evolution of GSM is via stepwise upgrades through the extension and the further development of existing cellular systems. This provides a faster availability of service and the opportunity to prepare customers for new services, 2.5G or 3G. GPRS or 2.5G can be built on the top of the existing GSM infrastructures; thus, the existing infrastructure and equipment can be reused. In other words, the GSM base station subsystem (BTS and BSC) is unchanged. A packet-based switching subsystem is being added to the GSM core subsystem. 3G has a high data rate and requires more cells. More GSM base station subsystems have to be added. There will be significant investment in new equipment and software.

# THE WIRELESS PROTOCOLS
## World Wide Web (WWW) Model

The Internet WWW architecture consists of clients (browsers), servers (web servers), and the wired network. Here, it is called the wired Internet. HyperText Transfer Protocol (HTTP) stands at the core of the World Wide Web. The HTTP controls the process between client and server. The communication is handled through TCP/IP (transmission control protocol/Internet

*Figure 3: The Wired and Mobile Web Server*

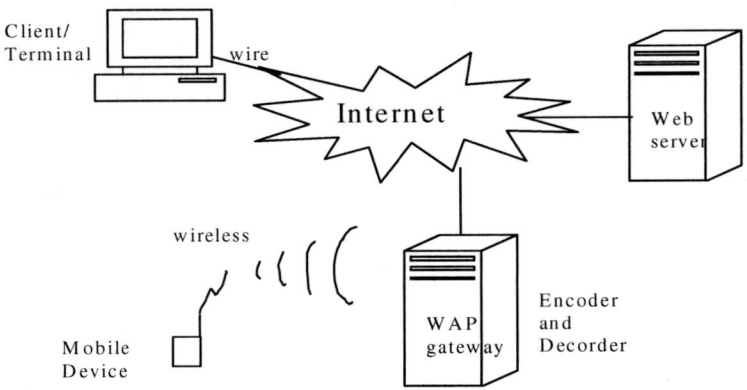

protocol). The Web server is for storing content, i.e., Web pages written in a special language, hypertext markup language (HTML). To access Web pages, the client sends a request in the form of a Uniform Resource Locator (URL). In response, the Web server sends the HTML pages to the client. Then, the client uses a browser to view the content.

Compared to the wired Internet, the wireless network is characterized by limited bandwidth, high latencies, high data loss, and variation in long-term connectivity/availability. TCP/IP-based protocols work well over wireless connections, but the performance is slow. Additionally, most mobile devices have limited display capabilities and cannot deal with all the features of full HTML. Hence, many alternative protocols and markup languages were created for mobile environments. These are as follows:

- WAP (Wireless Application Protocol) with WML (Wireless Markup Language).
- i-mode® with compact HTML (cHTML).
- Palm Web Clipping solution with Web Clipping Application format.

## The WAP Protocol

According to the WAP Forum's white paper, WAP (Wireless Application Protocol) is an open, global specification that empowers mobile users with wireless devices to easily access and interact with information and services instantly (WAP Forum Ltd., 1999). The WAP is a standard for providing cellular phones, pagers, and other handheld devices with secure access to e-mail and text-based Web pages. Introduced in 1997 by Unwired Planet (now Openwave), Ericsson, Motorola, and Nokia, the protocol is designed to make use of the existing Web technologies. It also consists of a client but with microbrowser or mobile browser, Web server, and an additional WAP gateway. The WAP gateway acts as a translator between the client microbrowser and the Web server. The connection between the Web server and the WAP gateway is wired, but the network between the WAP gateway and the client is wireless. Shown in Figure 3 are the wired and mobile Web servers.

The communication between the WAP gateway and the client is binary encoded to reduce network traffic. The WAP gateway encodes and decodes all messages respectively. Unlike the Web page of WWW, WAP is based on a "deck of cards" metaphor that contains content and script programming (Hjelm, 2000). Thus, a deck of cards can be sent to the client, and the order of cards can be based on user input. The markup language is called WML. It also uses WMLScript, a compact JavaScript-like language that runs with limited memory. The existing Web server is fully capable of delivering WAP content after some configuration change in Web-server software. A WAP gateway is required to sit between the wireless client and the Web server that converts WAP requests into HTTP and vice versa. WAP runs over all the major wireless networks. It is device independent, requiring only a minimum functionality in the unit so that it can be used with all enabled mobile phones and handheld devices. To build a wireless Web server, we need to add a WAP gateway to the existing Web server. Hence:

*Wireless Web server = Wired Web server + WAP gateway*

The design objective of the gateway is to tackle the low bandwidth in the wireless network connection. The WAP gateway performs mainly four functions: switching between transport protocols, compression, compilation, and decompression. However, the crucial weakness in WAP protocol is that all data transferred between the WAP client and the Web server is decrypted at the WAP gateway, i.e., all data, such as credit card numbers, etc., exist as free text in the memory of the gateway (Christian & Jogensen, 2001). The WAP protocol fails to provide end-to-end security, i.e., from client browser to Web server. End users lose direct control of the privacy of the data. WAP 2.0 addressed this problem by proposing alternatives, such as discarding the WAP gateway, placing the WAP gateway at the Web server end of the connection, or using application level security on top of WAP. However, there is a trade off between optimization and security. On the other hand, i-mode and Web clipping can provide end-to-end security. Both are proprietary systems.

## i-Mode

The i-mode is a wireless service developed by NTT DoCoMo in Japan. It is designed to provide mobile phone voice service, Internet, and e-mail. The i-mode protocol uses cHTML as its markup language. NTT DoCoMo constructed a packet-switched network alongside existing digital cellular networks that allows for a constant connection with the Internet (Ratliff, 2000). The architecture of i-mode is shown in Figure 4. When any developer designs secure i-mode applications, NTT DoCoMo requires the developer to sign a confidentiality agreement in order to access the i-mode proprietary Application Program Interfaces (APIs). It is predominately used in Japan. Also, each carrier in Japan has an Official Site Service. The content provider (Web site or commerce site) needs to register with the carrier to become an Official Site, which will be listed on the home page of the carrier. It is difficult to type long URLs on a mobile phone. The user can connect to the Official Site easily by navigating the menu, thus giving the content provider a much higher chance of having the site accessed.

*Figure 4: The Architecture of i-Mode*

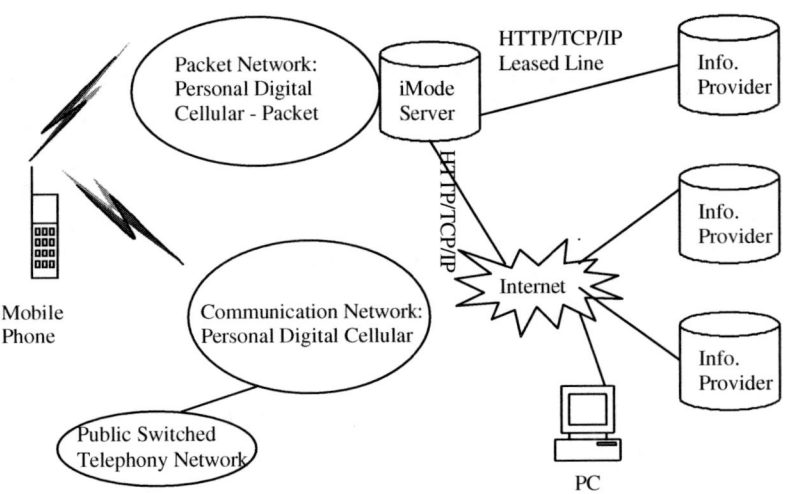

Another benefit of the Official Site is that the carrier will collect access fees from users when they access the content. Having payment services provided by the carrier makes it easy for content providers to engage in m-commerce without having to address issues related to payment systems and technologies. NTT DoCoMo includes a 9% surcharge on all fees collected for the content providers. Furthermore, the Official Site is the security. The i-mode provides a dedicated line connection service between the service provider i-mode gateway and the Web server of the Official Site. Content providers, such as financial institutions, have a highly secure connection. The carrier can use the mobile phone unique identifier and the IP address of the carrier gateway to build a secure system.

## Web Clipping

Web Clipping is a proprietary technology developed by Palm. The architecture of Web Clipping comprises the Palm device, a Web-Clipping proxy server, and the content server. Clipper software is installed in the Palm device to support Web Clipping. It is able to interpret and present a proprietary format called Web Clipping application (WCA) format. The format reduces the amount of data transferred over the air and the required storage on the Palm. The Web Clipping proxy is for translating the HTML content into this proprietary WCA format, and it resides at the 3Com Corporation's data center. Encryption and authentication between the handheld and the Web-Clipping proxy server is performed by Elliptic Curve Cryptography, which offers an extremely high level of security with a small amount of code. As a result, the data is encrypted and transported directly to the enterprise or content server, passing through any network as if it was in a sealed envelope.

## WAP or i-Mode

Among worldwide mobile Internet users, only 1% use Palm, 61% use i-mode, and 38% use WAP (Vacca, 2002). WAP and i-mode are the popular standards for writing m-commerce applications. Since the introduction of i-mode in February 1999, there are 30 million users today. Almost everyone is active in the service. On the other hand, the launch of the WAP service in Europe failed, as the services were not as good as users expected—they were slow and not easy to navigate using a small-sized phone. The i-mode service is a huge success. NTT DoCoMo is forming strategic global partnerships to implement i-mode services; they teamed up with AT&T in the United States and Hutchison in the United Kingdom. Summing up, there are several factors that contributed to this success:

1.  The wireless protocol is simple. It is a reduced version of HTML, called cHTML, which technically enables users to access desktop HTML. In other words, i-mode enables users to access any Web site.
2.  The connection cost is cheap. It is based on the data and not on the connection time. In addition, the price of the mobile phone with a big screen (full color with 10 lines of text) is affordable.
3.  The wireless technology used is packet switching. The service is always on whenever the mobile phone is on, though the data rate is only 9600 bps.
4.  Industrial cooperation played an important role. There is a close relationship between the hardware manufacturers and mobile phone operators. The phone manufacturers always meet the mobile phone operators' specifications.
5.  The service is personalized and fun-oriented and is targeted at people in their 20s, who are familiar with small devices such as Gameboy and minidisk.
6.  Last, it is geography and culture. About 50% of users are from Tokyo, where space is a scarce resource. A desktop computer occupies a lot of space. In addition, a fixed line to access the Internet is very expensive. A desktop PC to surf the Internet is a luxury.

The WAP service was launched in Europe almost at the same time, but the response was poor. Many say that it is a flop. The reasons for the failure are as follows:

1.  WAP protocol is a cumbersome language, which rewrites the protocol of TCP/IP that is unnecessarily complex. WAP cannot access Web sites that are written in HTML, but i-mode can.
2.  The cost of connection is much more than the normal fixed-line phone. In addition, a color WAP-enabled handset is very expensive (costs around US$600). A cheap WAP phone costs only about US$100. Furthermore, there was a shortage of WAP phones during the introduction phase.
3.  The wireless technology is circuit switching. It is not in always-on mode. Users have to go through a lengthy dial-up procedure from time to time.

Though WAP is an open standard, it has become clear that i-mode, the proprietary standard, will play a dominant role in m-commerce applications. As

2.5G and 3G will be widely available, the data throughputs will be high. In addition, improved technology enhanced the performance of the handheld devices. The size of the RAM more than doubled every year, and the processor speed increased every three to six months (Zetie, 2002). The growth rate of technologies such as compact flash memory type II offers 1 Gbytes or more. Toshiba has a 10 Gbytes drive in PCMCIA format. The HTTP protocol and markup language are an attractive solution for mobile devices.

## Short Message Service

The Short Message Service (SMS) has grown rapidly and is popular in Europe. SMS messages are two-way alphanumeric paging messages up to 160 characters that can be sent to and from mobile phones. SMS messages are transmitted over the mobile phone's air interface, using the signaling channels. Therefore, there is no delay for call setup. SMS messages are stored in the Short Message Service Center and sent to the recipient when the subscriber connects to the network. According to new figures from the Mobile Data Association, mobile users in the United Kingdom sent an average of 2 million SMS messages per hour during September 2002. The popularity is due to the low cost, around 10p (US$0.015) for sending a message, while voice costs 9p and 30p per minute for land-line phone and mobile phone, respectively, at peak times. SMS generated significant revenues not only for network operators but also for Web portals. In the United States, however, the growth of SMS over mobile phones is hindered by a lack of carrier interoperability.

# CHALLENGES IN M-COMMERCE

NTT DoCoMo released 3G in May 2001. The response is unexpectedly low. Only 114,500 users signed up for the 3G services. This failure is due to the high cost of 3G services. Also, the users have to buy a new mobile phone. Furthermore, the networks have limited coverage. M-commerce, which is more complex than e-commerce, faces a number of challenges (see Figure 5).

## Delay in 3G

Mobile network operators (MNOs) are pushing m-commerce. The success of m-commerce in Japan changes the concept of "free" Internet to "paid"

*Figure 5: Challenges in M-Commerce*

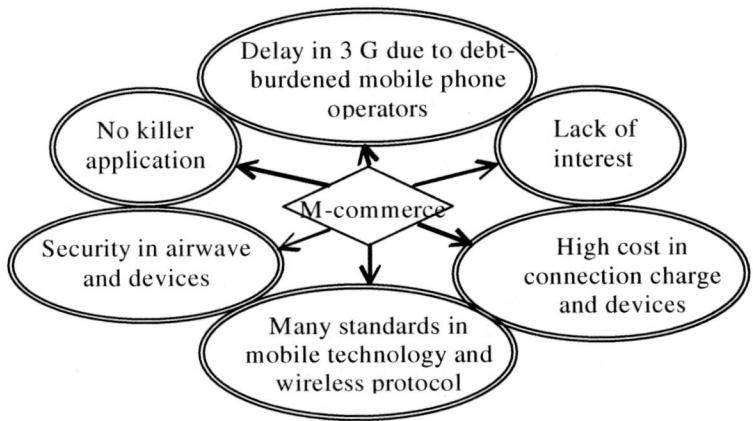

Internet. Users are willing to pay for the service. MNOs have foreseen a huge profit in taking control of the backbone of m-commerce—the wireless infrastructure. In addition, MNOs also play a dominant position in providing m-commerce applications. This created unreasonably high expectations of 3G services. Big companies in Europe, such as Deutsche Telecom, France Télécom, Spain's Telefónica, and the U.K.'s Vodafone spent an estimated US$125 billion to US$150 billion on 3G licenses (Garber, 2002). Many of them are burdened with high debts. Banks downgraded their credit ratings. At the same time, the global economy is in a downturn after September 11, 2002, and the accounting scandal at Enron, the energy firm, which was followed by similar problems at WorldCom, the telecommunication giant in the United States. All of this bad news destroyed the confidence of investors in technology stock. It is extremely difficult for any of the MNOs to raise funds through the capital markets. Hutchison postponed the plan of raising money through the issue of bonds. With their huge debts, network operators find it hard to cut costs. Spectrum trading and network sharing are the ways to cut costs. These enable network operators to save money when deploying 3G. However, there are many unknowns in sharing and spectrum trading: competition, ownership of the license obligations, and responsibility for the license debts (Webb, 2002). Government support and assistance are important in resolving difficulties that arise.

The 3G requires more cells, and thus, more new radio stations have to be built. This requires a lot of investment. With huge debts, MNOs have significant difficulties in meeting the requirements to provide the new 3G services. MobilCom, the German mobile operator, is cutting more than half of its workforce. The preparation for 3G is on the shelf. In Singapore, mobile phone operators held discussions with governments to delay the rollout deadline for 3G. The debut of integrated 3G IC solutions was delayed; the "life-cycle" of GSM, CDMA, TDMA, AMPS, and particularly, 2.5GPRS systems will be extended (Telford, 2002). In the near term, the future of 3G looks gloomy. NTT DoCoMo wrote off a 126 billion yen (about US$1 billion) investment in Hutchison 3G (U.K.). Also, Orange, which is billions of euros in debt, shelved its plan to run a 3G mobile service in Sweden.

## Lack of Interest and High Cost

On the other hand, MNOs in Europe have the problem of falling revenues. According to Gartner's report, the connection growth for last year declined from 57.4% to a miserable 19.5% in the United Kingdom (NUA analysis, 2002). The Western European market reached the saturation point, where the mobile possession rate is close to 100% in some countries. In addition, mobile users have "upgrade fatigue," i.e., they are reluctant to upgrade their mobile phones. A report from the World Market Research Center, released in May 2002, predicted that 3G sales are likely to be sluggish because of a lack of consumer interest in new services. The research done by the mobile consultancy group Detica also warned that there is no single "3G killer application" that will entice large numbers of people to embrace 3G. In 2002, the mobile phone business pushed hard on picture messaging, which requires new expensive handsets. The response has been poor. The mobile revenue mainly comes from voice calls and SMS messaging.

## Young Market with Many Standards

This is a young market with many existing standards. 3G standards have some commonalities, but they are not fully compatible with each other at the air interface (radio-transmission) level. This imposes a limitation on global roaming and the creation of mass-market volumes for terminals and infrastructures. The third-generation partnership project (3GPP) was formed to produce technical specifications for 3G mobile systems and works with different 3G systems. In addition, the spectrum problem in the United States casts a shadow on global roaming. The Department of Defense occupies the spectrum band needed for 3G services. After September 11, 2002, the U.S. military is unlikely to give up the spectrum. This introduces further technical difficulties. In addition, Japan is going it alone to develop a fourth-generation standard (4G) that could offer data rates of up to 100 Mbits/s. However, during an International Telecommunication Union meeting on 4G, Japanese representatives limited the debate in order to enshrine Japan's vision as the de facto standard for 4G (Wireless Web Analysis, 2002).

WAP is disappointing. Though i-mode is popular, it is proprietary. However, both formats have the quality that they benefit from each other. There is a need for an open standard for wireless protocol. Industry cooperation is a must if a solution is to be found. At present, industry is working toward limiting the number of markup languages available for mobile devices. Both WML and cHTML are now migrating into XHTML (extensible HTML), which are XML based. There are still many multiple proprietary browsers and operating systems. Developers find it hard to develop applications with the myriad of handheld devices. In the future, it all depends on the role of XML and the evolution of the WAP standard.

The advances in technology resulted in a convergence of the functions of mobile phone and PDA devices. Mobile phones are commonly equipped with PDA functionality, such as a large screen and easier methods of input, e.g., pen with touch screen. PDAs have a phone function. There are different kinds of handheld devices from phone and computer manufacturers. The market for handheld devices is different from the personal computer market. For instance, Nokia not only produces the phone (hardware) but also develops Symbian (the software), together with other phone manufacturers such as Motorola. This is also the problem with PDAs. Each device comes with its own proprietary operating system. Many standards are competing with each other. Microsoft joined the competition by rolling out two operating systems: Pocket PC2002 and the smartphone system for PDAs and mobile phones, respectively, in 2002. Both systems can be synchronized to a desktop PC, a convenience for business users. Microsoft, together with Orange, launched an SPV (Sound, Picture, and Video) phone, the first mobile phone with the Microsoft smartphone system, in October 2002. The participation of Microsoft in the handheld device market will have a major impact on market development.

## Security

Mobile communications offer users many benefits, such as portability, flexibility, and increased productivity. Handheld devices allow remote users to synchronize personal databases, and they provide access to e-mail, Web browsing, and Internet access. The most significant difference between wired networks and mobile communication is that the airwave is openly exposed to intruders. This is the main source of risk—some existing problems have been exacerbated and some are new. Analog phones using 1G technologies are more vulnerable to eavesdropping than digital cell phones. The unencrypted analog signal can be intercepted by using simple radio scanners, whereas digital phones are protected using encryption. Nevertheless, digital systems are also subject to attack. For instance, there is the "man-in-the-middle" problem, where the intruder sets up a base station transceiver with a modified phone. The intruders put themselves in between the target user and a genuine network and have the ability to eavesdrop, modify, delete, reorder, replay, and send spoof signaling. The intruders eavesdrop on signaling and data connections associated with other users by using a modified mobile phone. 3G imposes more security measures to defeat false base-station attacks. A security mechanism including a number sequence is implemented to ensure that the mobile can identify the network. Overall, 3G technology has far greater security than 2G and 2.5G.

The wireless handheld device provides the advantage of mobility for mobile workers but also creates new types of security risks. The types of risks are summarized as follows (Karygiannis & Owens, 2002):

1.  Their small size, relatively low cost, and constant mobility make them likely to be lost or stolen. Sensitive information such as the "private-key" is thus vulnerable. The mobile phone tops the list of property left behind in the U.K. railway system.
2.  PDAs can beam information from an IR (Infrared) port to another IR port to easily exchange contact information such as telephone numbers. The data is unencrypted, and any user who is in close proximity to the handheld device and has an appropriate device pointed in the right direction can intercept and read the data. This is known as data leakage.
3.  Smart phones may lose network connectively not only when they are outside the coverage area but also from the use of cell phone jammers, which block cell phone communication. Many restaurant and movie theaters installed such devices without warning.
4.  A 3G mobile device, when connected to an IP network, is in the "always-on" mode. This mode minimizes the need for the device to go through an authentication procedure each time a network request is made. Thus, this

"always-on" mode makes the device susceptible to attack. Moreover, it also provides the opportunity to track a user's activity, which may be a violation of privacy.

Viruses have not been widely considered a security threat in PDAs or mobile phones because of their limited memory and processing power. Moreover, users typically synchronize their data with their PCs, so they can recover any lost or corrupted data simply by synchronizing with their PCs.

## Limitations of Handheld Devices

Technological developments will increase the computing power and storage in handheld devices. However, insufficient battery life and power consumption will impede the potential growth of m-commerce, even when 3G is widely available. At present, the battery life is very short. It can last for two or three hours for surfing the Internet or talking. The battery has to be recharged from time to time, or a second battery must be used. Fuel cells will be a solution to this problem, but they are in the early stages of introduction and will be expensive. In the near future, short battery life is a major barrier in handheld devices. Besides this, the small screen is another limitation. PDAs have a larger screen (4.9×3.1 inches). The color touch screen makes it easier for users to navigate the Web, however, they are expensive, normally around US$600 compared to a black and white one priced at US$100. Furthermore, mobile phone users still prefer the small-sized phones. The screen of a mobile phone is 2×2 inches, which causes difficulty when surfing the Web. Many mobile users are aware of 3G but do not understand the "user experience." A low-power, inexpensive, high-resolution color display would seriously increase the growth of m-commerce.

# M-COMMERCE APPLICATIONS

The mobile telecommunications market reached critical mass in certain areas and is growing rapidly in others. To study the phenomenal growth, we examine the mobile market penetration rate in three areas: Japan, Europe, and the United States. In Japan, the number of mobile phone Internet users has reached 35 million. This implies that one in four people can connect to the Internet by mobile phones. In Europe, most of the countries experienced a two-

*Figure 6: M-Commerce Applications*

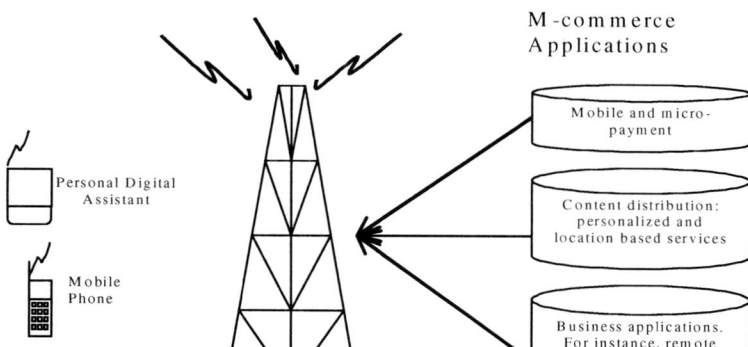

digit growth rate from 1999-2001. In the United States, there is no dominant technology for mobile devices. The possession rate of mobile phones is around 30%. And, less than 3% of users access the Internet via handheld devices.

E-commerce is characterized by online bargain buying, whereas m-commerce is personalized and ideal for access to location-based services. Many new business models were established around the use of mobile devices. Mobile devices have the characteristics of portability, low cost, more personalization, Global Positioning System (GPS), voice, and so on. The new business models include micropayment and mobile payment, content distribution services, and business services. Illustrated in Figure 6 are m-commerce applications. Because of their existing customer base, technical expertise, and familiarity with billing, mobile telephone operators are the natural candidates for the provision of mobile and micropayment services. Micropayment involves small purchases, such as vending and other items. A mobile device can communicate with a vending machine using a local wireless network to purchase desired items. In other words, the mobile phone is used as an ATM card or debit card. It is on trial in Japan by NTT DoCoMo with Coca-Cola and Itochu. The mobile operator NTT DoCoMo has 30 million users. Mobile users will be able to make purchases from Coke vending machines on their mobile phones. Later this year in Europe, Vodafone, the largest operator in Europe with 100 million users, will also roll out its payment platform—allowing customers to make purchases with their mobile devices seamlessly across most of Europe. Trial customers will be able to purchase goods over the Internet and in retail shops using their mobile devices. Eventually, Vodafone will extend the service to automatic purchase points, such as vending machines. This can be done for small purchases, however, for large transactions, banks must be involved. In France, Mastercard, together with Oberthur Smart Cards, France Telecom, Europay, and Motorola, is to run a pilot scheme to conduct wireless payments by inserting smart cards, which will be issued by Credit Mutuel, into dual-slot handsets.

Content distribution services are concerned with real-time information, notification, and a positioning system for personalized information delivery related to location. Real-time information, such as news, traffic reports, stock prices, and weather forecasts, can be distributed to mobile phones via the Internet. The information is personalized to the user's interests. As everyone is interested in his or her own wealth, stock quotes have the highest usage, with e-mail being second in usage (Chae et al., 2000). Notification can also be sent to the mobile device. For instance, a dot.com's chief executive is provided with the current reading of how many hits his or her site had. By using a positioning system, you can retrieve local information, such as restaurants, traffic, and shopping information. Mobile devices offer advertisers a high rate of response, as advertisements are personal. SMS is a good advertising tool. After piloting two SMS marketing campaigns in different countries, Pepsi found that SMS marketing is an effective advertising medium, especially for target customers (Mobileinfo, 2002). Entertainment and games are also important markets for m-commerce. The content distribution services with a greater degree of personalization and localization can be effectively provided through a mobile portal. Localization means to supply information relevant to the current location of the user. Users' profiles, such as past behaviours, situations, and locations, should be taken into account for personalization and localized service provision. MNOs have a number of advantages over the other portal players (Tsalgatidou & Veijalainen, 2000). First, they have an existing customer relationship and can identify the location of the subscriber. Second, they have a billing relationship with the customers, while the traditional portal does not. MNOs can act as a trusted third party and play a dominant role in m-commerce applications.

Businesses also need to think across traditional boundaries in m-commerce. Interaction among businesses, consumers, and smart appliances creates a lot of new opportunities. First, appliance-to-appliance, that is, appliances interact with an automatic service scheduler. Second, appliance-to-business, which can deliver smart appliance repair alerts. Third, business-to-appliance can deliver remote activation of smart appliances. For instance, the built-in sensors in a car will inform the repair service as to what part is broken. Many automotive makers are planning to implement Telematics providing two-way communication between car and a service center via the mobile phone networks and the Internet. In the near future, in the case of car breakdown, the command center staff will communicate with the car's engine management system and repair the vehicle remotely. The key is to develop services that are sufficiently critical that people will be willing to pay for them. It is important to cooperate with other partners in rolling out m-commerce solutions.

M-commerce also has a great impact on business applications, especially for companies with remote staff. Extending the existing Enterprise Resource Planning (ERP) systems with mobile functionality will provide remote staff, such as sales personnel, with real-time corporate and management data. Time and location constraints are reduced, and the ability of mobile employees is enhanced. For instance, businesses such as utility services can send service calls to engineers in the field. On the other hand, the engineers can send back service reports and access back-office systems to order parts, download electronic manuals for fixing appliances, and so on. Sales representatives can place an order and check stock levels with PDAs, too. As a result, real-time data is available for planning daily operations. The logistic-related business also benefits from the use of mobile inventory management applications. One interesting application is "rolling inventory" (Varshney & Vetter, 2002). For this scenario, multiple trucks carry a large amount of inventory while on the move. Whenever a store needs certain items/goods, a nearby truck can be located, and just-in-time delivery of goods can be performed. M-commerce offers tremendous potential for businesses to respond quickly in supply chains.

Shopping is a part of everyday life. E-commerce, which is characterized by bargain buying, can reach out to a global market, i.e., a macromarket level, with a critical mass, instantly. On the other hand, m-commerce is characterized by location and personal preference and is affected by the culture and environment of the target users. M-commerce can focus on a micromarket level. The application of mobile and micropayment will be successful in those areas where electronic payment is prevalent, and people are used to electronic money rather than real money. In Hong Kong, smart card payment is being used in different kinds of transportation systems, such as bus, underground, and ferry. The smart card can be topped up in any 24-hour convenience shop. This payment culture is compatible with micropayment technology. Because the penetration rate of mobile phones is around 90%, users will welcome micropayment technology. In the United Kingdom, because transport is still partly paid for using real money, the implementation of it will need a change in the behavior of people. Concerted coordination is required among the participating parties.

In m-commerce, most of the content is charged for. Because it is a paid-for service, cost will play a critical factor in its success. For instance, mobile phone users are charged only US$0.8 a month to have news delivered in Japan. This attracted millions of users. Besides bargain shopping, the Internet is amassing information about health care, government services, and potential purchases. Existing Web portals will naturally become the content distributors. The Web portals now have the problem of declining revenue from advertising, their main income. Distributing personal content to mobile users provides them with an opportunity to diversify their sources of revenue. The success of content distribution will not only depend on the cost but also on the kinds of information that appeal to individual users.

Security is the main concern of business to adapt m-commerce for their intranet and extranet applications. If m-commerce has the same security level as wired e-commerce, business will use it to speed up the business process. In addition, the high cost of handheld devices and the lack of standards in software and hardware deter its adoption by business. The integration of m-commerce with e-commerce will certainly increase a firm's competitive advantage by providing quick response to customers—both external and internal. Business has to carefully examine the potential offered by m-commerce.

# ISSUES IN IMPLEMENTING M-COMMERCE

The mobile Internet is ideal for particular applications and has useful characteristics that offer a range of services and content. As the connection charge and the handset are still costly, this caters for a niche market of mobile users who are young and affluent and mobile business users. For example, Tesco's mobile service is targeted toward business executives. It offers convenience for those who use trains or public transport to get home. They can place orders on their way home. When they get off the train, they just go to the nearby Tesco store to pick up the order; this saves a lot of shopping time. In the United Kingdom, the HSBC bank notifies customers about their current balance by SMS messaging. It is a free service and is welcomed by many mobile users. Utility services such as British Gas use mobiles extensively. The off-site engineers access back-office systems to order parts, download schematics for fixing appliances, and so on, through handheld devices. So when implementing m-commerce, businesses should consider the following issues:

1. Understand the position of the business in the industry.
2. Evaluate the overall usefulness and convenience to the customers—both internal, e.g., off-site employees, and external.
3. Develop applications with a fun and hassle-free user interface.
4. Offer fast and secure ordering facilities.
5. Allow users to recover from an aborted session.

The advent of m-commerce is fast approaching. In business, the constant word is "change." Winning customers in today's highly competitive, demanding world is the key to survival. The mobile world is changing the logic of business, businesses have to implement effective strategies to capture and retain increasingly demanding and sophisticated customers. Furthermore, it is timely for the business community to prepare for the approaching economic boom, which always follows an economic downturn. Business needs to think critically about how to integrate the mobile Web to the wired Web, which requires careful strategic thought, as not all applications are appropriate. Those that make the correct decisions will control the future.

## ACKNOWLEDGMENTS

The authors wish to acknowledge Kwan-Ming Wan, IBM UK, for consultation and Dr. Qian Wang, Tareg Saad, and Ioannis I. Kypraios for their valuable information and fruitful discussions. We are also grateful to the editor of this book, Dr. Nansi Shi, for his comments.

## REFERENCES

Abbott, L. (2002). *M-commerce*. Available at: http://www.mobileinfo.com/Mcommerce/driving_factors.html.
Chai, M., Choi, Y., Kim, H., Yu, H., & Kim, J. (2000). Premier PAS of mobile Internet business: A survey research on mobile Internet service. In *Proceedings of International Conference of KMIS/OA*, Seoul, Korea.
Christian, N. & Jorgensen, N. (2001). WAP may stumble over the Gateway (security in WAP-based Mobile Commerce). In *Proceedings of SSGRR*.

Garber, L. (2002). Will 3G *really* be the next big wireless technology? *Computer, IEEE,* 35(1), 26-32.

Hjelm, J. (2000). *Designing Wireless Information Services.* New York: John Wiley & Sons.

Karygiannis, T. & Owens, L. (2002). Draft: Wireless Network Security: 802.11, Bluetooth™ and the Handheld Devices. National Institute of Standards and Technology, Technology Administration, U.S. Department of Commerce, Special Publication 800-48.

Mobileinfo. (2002). *Pepsi exploits SMS to promote soft drink.* Available at: http://mobileinfor.com/News_2002/Issues 18/Pepsi_SMS.htm.

NUA analysis. (2002). *3G or not 3G.* July 1, NUA Internet Surveys. Available at: http://www.nua.ie/surveys/.

Ratliff, J. (2000). *DoCoMo as national champion: i-mode, W-CDMA and NTT's role as Japan's pilot organization in global telecommunications.* Working paper, Department of Sociology, Santa Clara University, California, USA.

Shirky, C. (2002). Sorry wrong number: The world's favorite statistics about phone use is no longer true. So why won't people just drop it? *Wired,* (October), 97-98.

Telford, M. (2001). Trends in mobile telecom. *The Advanced Semiconductor Magazine, September,* 14(7), 32-37.

Tsalgatidou, A. & Veijalainen, J. (2000). Mobile electronic commerce: Emerging issues. In *Proceedings of EC-WEB 2000, First International Conference on E-commerce and Web Technologies,* London, Greenwich, UK, (September 2000), *Lecture Notes in Computer Science, 1875* (pp. 477-486). Heidelberg: Springer Verlag.

UMTS Forum. (2002). White Paper No. 1. Evolution to 3G/UMTS Services. Available at: http://www.unts-forum.org/reports.html.

Vacca, J. R. (2002). *I-mode Crash Course.* New York: McGraw-Hill.

Varshney, U. & Vetter, R. (2002). Mobile commerce: Framework, applications and networking support. *Mobile Networks and Applications,* 7, 185-198.

Walke, B. H. (2002). *Mobile Radio Networks: Networking, Protocols and Traffic Performance* (second ed.). New York: John Wiley & Sons.

Webb, W. (2002). Spectrum trading & network sharing could enable operators to save money when deploying 3G. *Wireless Web Analysis.* Available at: http://wireless.iop.org/article/news/3/10/1.

Wireless Application Protocol Forum Ltd. (1999). WAP white paper. Available at: http://www.wapforum.org.

Wireless Web Analysis. (2002). Japan sets the pace for 4G technology. Retrieved October 3. Available at: http://wireless.iop.org/article/news/3/10/2.

Zetie, C. (2002). PDAs: Always pushing the barrier. *InformationWeek*, (August 12). Available at: http://www.informationweek.com.

# Section II:

# Mobile Business Network and Consumers

**Chapter V**

# Understanding Emergent M-Commerce Services by Using Business Network Analysis: The Case of Finland

Tommi Pelkonen, Helsinki School of Economics, Finland

Nikhilesh Dholakia, University of Rhode Island, USA

## ABSTRACT

*Successful m-commerce business models depend on complex networks of business relationships, comprising telecommunications service providers, mobile device makers, financial linkage providers, and various third-party value-adding companies. In this chapter, we will discuss such business relationship networks in the context of Finland, and offer general guidance for the formation and sustenance of effective business networks for m-commerce players worldwide.*

# M-COMMERCE AND BUSINESS NETWORKS

Across the globe, mobile commerce (m-commerce) service providers are testing or planning to test a wide variety of business models. One of the striking features of m-commerce business models is the complex network of business relationships needed to create, launch, and sustain such services. Such business networks are comprised of telecommunications service providers, various types of device makers, payment systems and financial linkage providers, and various third-party value-adding companies.

A key question for m-commerce strategists and their financial supporters is this: What types of business network arrangements work toward promoting early success and long-term sustainability of m-commerce ventures? To seek an answer to this question, it is helpful to do the following:

*   Focus on a lead country where mobile telecommunications and m-commerce have developed further than most other places.
*   Draw from theoretical schemas that help us to understand complex global networks, especially networks that cross the boundaries of many nations.

From the mid-1980s, Finland has been one the leading countries in developing and deploying mobile services, in terms of per capita availability of mobile handset terminals and mobile service accounts. With its flagship company Nokia, Finland is a global technological leader in the development of mobile communication networks and terminals. By the start of the new millennium, large numbers of new start-up companies emerged in Finland to serve Nokia's and other companies' needs for mobile applications and service development. Regions such as the Helsinki suburb of Espoo and the remote Arctic Circle city of Oulu have developed as mini-Silicon Valleys, with many startup firms focused strongly on mobile communications and m-commerce.

The number and diversity of agreements, strategic alliances, and mergers featuring firms from Finland and crisscrossing geographical barriers has been staggering. The merger of Sweden's Telia and Finland's Sonera adds to this growing list. This merger represents a step toward the emergence of a pan-Nordic/Baltic telecommunications operator, with the requisite critical mass needed to thrive in that region as well as to make some global impact. Just as SAS—based on the combined strengths of Sweden, Denmark, and Norway—succeeded in the global airline market, the newly merged Scandinavian operator will have a global footprint.

Business network theories enjoyed decades of popularity in Scandinavia and were employed to study the internationalization and global linkages of

Scandinavian firms. Presented in this chapter is an approach to understanding the mobile telecommunications industry, and especially, the emergent m-commerce space in Finland, using the Scandinavian-inspired business network analysis methods. In this chapter, we aim to illustrate the dynamics of the industry by presenting a market model based on business network theories. The information in this chapter is drawn from studies conducted at the Helsinki School of Economics as well as the experience of the authors.

While the chapter draws mainly from the Finnish experience, the overall objective is to create a generic analysis tool for m-commerce market actors to use to assess their strategic positions. The key questions addressed are the following:

1.　What kinds of actors are there in the mobile telecommunication business, and especially in value-added m-commerce services?
2.　What kinds of resources do these actors rely on?
3.　How are the various actors related to each other?
4.　How do these relationships help or hinder the strategy of each actor?
5.　What general lessons can be drawn about the potential successes and failures of m-commerce actors and strategies?

Developed in this chapter first is a framework from illustrative analyses of the Finnish situation. Next, the model is outlined in a generic form. Suggestions are then provided for customization of the model for various countries and different market needs. Conclusions and future research directions round out the chapter.

# FINLAND AS A
# MOBILE BUSINESS PIONEER

The Finnish telecom industry dates back to 19th century, when the first telecom networks were built in the country, initiated by the Russian Czar and the Finnish autonomous government. Until the 1980s, because of limited capital resources available in the country, local telecom companies and cooperatives dominated the Finnish telecommunications markets. Yet, the nearly 300 local telephone companies had talented personnel, and Finland's telecom industry, led by the national monopoly company Telecom Finland,[1] gradually developed one of the most sophisticated networks in Western Europe. Liberalization and digitization of the telecom network, strongly initiated and steered by the Finnish government, boosted the sophistication of the networks.

Telecommunications rose in importance for the Finnish economy throughout the 1990s. In Table 1, the rapid growth in importance of telecommunications for the economy is shown. Telecommunications represented only 2% of total Finnish GDP in 1990, but by 2000, this had doubled to about 4% of the economy. During the same period, mobile phone penetration rose from a mere 5% to over 70% of the Finnish population.

By the end of the 20th century, Finland became one of the leading countries in mobile communications. Nearly 78% of Finns had a mobile phone in 2001, and the market neared the point of saturation (Leppävuori, 2002). Among certain segments, e.g., teenagers and the business community, the penetration was nearly 100%. Public telephones have become obsolete, as nearly everyone has a personal portable phone. People moving into an apartment or a house often forego a fixed connection and subscribe only to the mobile network. The traditional fixed phone line is used mainly for high-speed Internet connections rather than for voice communications.[2]

The telecommunications sector also forms an essential part of the Finnish ICT-cluster.[3] Presented in Figure 1 is the structure of the various converging

*Table 1: Rapid Growth of Finnish Telecommunications Industry*

| Telecommunications Industry in Finland | 1990 | 1995 | 2000 |
|---|---|---|---|
| Total turnover of telecom companies, mEUR | 1428 | 1861 | 4364 |
| Telecom turnover as % of GDP | 2% | 3% | 4% |
| Personnel in the telecom industry | 20,067 | 16,405 | 24,204 |
| **Subscriptions** | **1990** | **1995** | **2000** |
| Fixed-network subscriptions, total | 2,670,000 | 2,810,000 | 2,848,000 |
| Fixed-network subscribers/100 inhabitants | 53.4 | 55.0 | 55.0 |
| Mobile-network subscriptions, total | 257,872 | 1,039,126 | 3,728,625 |
| Mobile-network subscribers/100 inhabitants | 5.2 | 20.4 | 72.0 |
| Public telephones, total | 20,229 | 25,267 | 12,427 |
| Public telephones/100 inhabitants | 0.4 | 0.5 | 0.2 |
| **Household information** | **1990** | **1995** | **2000** |
| Internet connections/100 households | — | — | 31.6 |
| Computers/100 households | — | — | 48.2 |
| Fixed-network subscriptions/100 households | 131.1 | 128.8 | 124.1 |
| Mobile phones/100 households | 12.7 | 47.6 | 162.5 |
| **Selected demographic information about Finland** | **1990** | **1995** | **2000** |
| Population | 4,998,000 | 5,117,000 | 5,176,000 |
| GNP, mEUR (market price) | 87,967 | 94,953 | 132,038 |
| Consumer price index (1995 = 100) | | 100 | 108 |

*Source: Ministry of Communications and Transport, Finland, 2002.*

industries, ranging from media to traditional telecommunications. Digital service companies constitute the core of the cluster. Digitization of information had dramatic impacts on end users and ICT companies. Digitization allowed ICT firms to start producing and offering innovative services that were totally new to end users, increased the geographical reach of existing digital and offline services, and offered cost-cutting opportunities for organizations adopting new digital technologies. Finnish companies are also actively building the digital communications infrastructure and developing new services.

In 1999, the total turnover of the Finnish ICT cluster was estimated to be 34 billion euros (nearly 25% of Finnish GDP), and the ICT-related activities employed around 146,000 people (about 6% of the total workforce) (Table 2).

In recent years, mobile telecommunications has been one of the most rapidly developing sectors within ICT. Finnish companies as well as users have been very open to modern technology, and Finland has become an interesting lead market for mobile development. By 2000, several venture capitalists began investing in Finnish start-up companies to explore and learn about possibilities in the mobile markets. In the next section, we look in detail at the structure of the Finnish mobile telecom markets.

*Figure 1: Generic Model of ICT-Cluster*

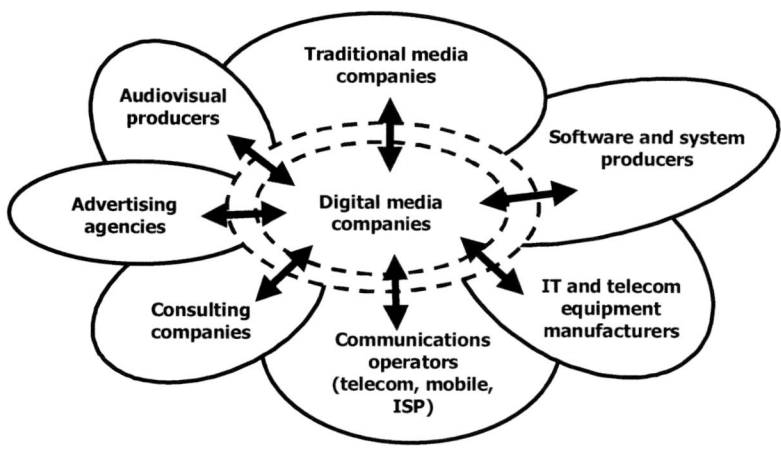

*Source: Kuokkanen, Toivola and Väänänen (1999).*

*Table 2: The Finnish ICT-Cluster 1999*

|  | Manufacturing | Service Creation | Communication Services | Content Production |
|---|---|---|---|---|
| **Turnover (in euros)** | • 16.3 billion | • 8.6 billion | • 3.7 billion | • 5.5 billion |
| **Personnel** | • 43,800 | • 42,000 | • 19,000 | • 41,000 |
| **Key products and services** | • Communication equipment<br>• Computers<br>• Consumer electronics<br>• Electronics components<br>• Measurement and automation equipment | • Software and system production<br>• Professional business services and consulting<br>• Wholesale of products | • Mobile communication services<br>• Fixed-network services (telecom- data and value-added services) | • Printed communications<br>• Electronic publishing<br>• Offline media<br>• TV and radio broadcasting<br>• Information services |
| **Total ICT cluster in Finland** | • Turnover 1999: about 34.1 billion EUR<br>• Personnel: 146,000 employees | | | |

*Source: Meristö, Leppimäki and Tammi (2002).*

# FINNISH M-COMMERCE: A BUSINESS NETWORK APPROACH

## Actors in the Finnish M-Commerce Space

Durlacher (2001) and Leppävuori (2002) described mobile telecom markets as special types of business networks called *value-webs*. They identified the core actor groups and their relationships. In this section, we illustrate a part of the value-web in Finland and then develop it more generically in later sections.

First, the Durlacher value-web divides mobile telecom actors into three main areas: (a) Services, (b) Applications, and (c) Technologies. Finnish telecom operator (e.g., Radiolinja, Sonera, DNA) and mobile portals (e.g., Sonera Zed) comprise the *service area* of the Finnish m-commerce value-web. The *application area* of the Finnish value-web consists of traditional content creators such as media companies (e.g., MTV3, Soneraplaza, Sanoma) and small start-ups creating or aggregating applications (e.g., SmallPlanet and MatchEm). Finland's leading technology company, Nokia, dominates the *technology* area of the country's m-commerce value-web. The giant is active in all three technology areas. Nokia's two core businesses—mobile handsets and network equipment—make this company a crucial partner for nearly all other Finnish mobile telecom actors. In Figure 2, the Durlacher approach to telecom value-webs is presented in more detail, with illustrative firms from the Finnish context.

*Figure 2: The Mobile Value-Web in Finland*

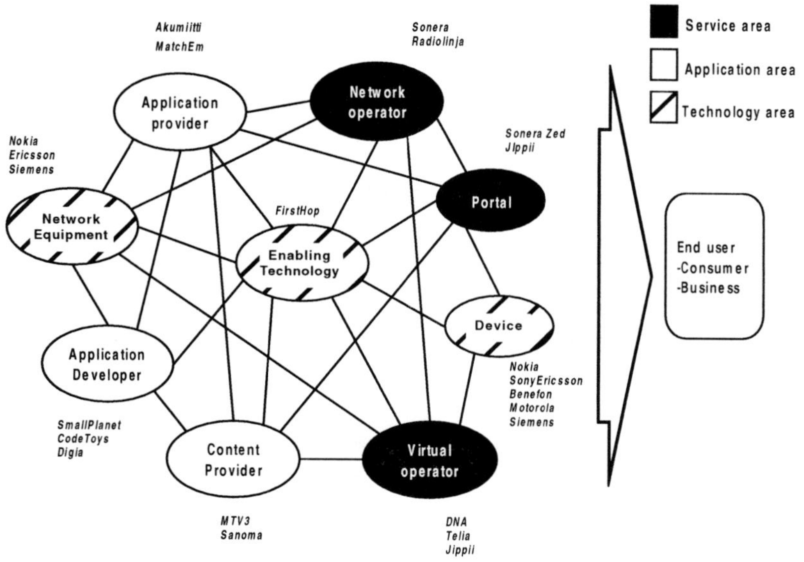

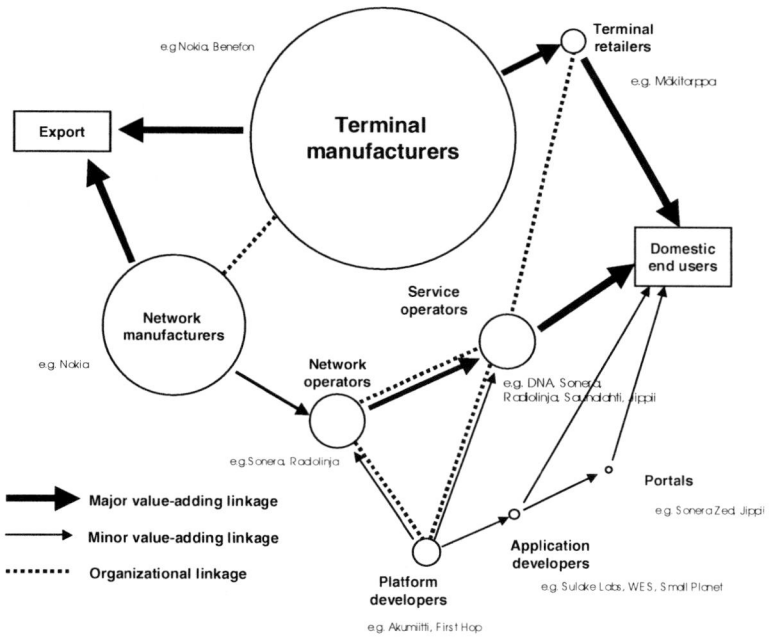

*Source: Modified from Eqvitec and Durlacher, 2001.*

*Figure 3: The Distance/Magnitude Value Network of the Core Actors in the Finnish Mobile Cluster*

*Source: Adapted from Leppävuori, 2002 (Enhanced by the Authors).*

Leppävuori utilized the Durlacher model to illustrate the power balance within the Finnish telecommunications landscape. The network model of Figure 3, as also the chart in Figure 4, presented later in this chapter, dramatically illustrate the strength of Nokia within the Finnish telecom markets. The application or platform development efforts of small start-up companies in mobile services have only a marginal impact on market development, indicated by the thin arrows in Figure 3. Telecom operators possess some market power, especially in relation to domestic end users. Accessing export markets, however, is challenging for both the telecom operators and mobile start-ups. Only Nokia has a strong global reach into export markets (Figure 3). Recently, the largest Finnish media companies took their initial steps in internationalization, e.g., Sanoma Magazines has become one of the leading European magazine publishers.[4]

## Resources
### Personnel

The telecom sector has long been dominated by technology and engineering skills. Finland is well known for its talented engineers, e.g., in the paper industry and machinery, and the Finnish telecom sector is no exception to this. In the mid-1990s, Finland's technical universities increased their student intake for telecom-related studies and were able to respond rapidly to the increasing human resources needs triggered by the constant international growth led by Nokia and its partners.

For Finnish mobile-telecom-related companies, a major challenge has been to find and retain people who have enough experience in international IT service business. Traditionally, Finnish organizations have been excellent at developing technology but short-staffed in experienced sales personnel. During the mobile boom toward the end of the 1990s, this shortage of experienced sales personnel struck hard. Venture capitalists financed innovative technology development, but only a few Finnish companies were able to create revenue on investment. Skill sets of the start-up companies were aimed at research and development, instead of at sustainable business development.

### Financial Resources

Finland has been one of the model countries for industrial growth since World War II. It transformed from mainly an agricultural economy in the early

1950s into one of the leading high-tech economies by the end of 1990s. Yet, the country's capital accumulation has been limited. Capital market liberalization in the 1980s opened the doors to foreign investment. Still, for most entrepreneurs, the only viable options for financing business development were through commitment of personal and family savings and bank loans. Risk-based venture funding and stock market financing existed only minimally. It took nearly 10 years for venture capitalists and business angels to really start investing in Finnish companies. At the peak of the hype, the first quarter of 2000, major financial investors were either actively investing in or actively seeking high-tech start-ups to invest in.

The sudden increase in venture capital flowing into the country created multiple side effects. Investors did not have the time to evaluate in detail the feasibility of the business models and capabilities of most start-up companies. Investments were made with minimal planning and high risk. For mobile start-ups, the situation was lucrative. With a credible "mobile story," substantial capital could be attracted to support the establishment of new businesses and to commit to expenditures for rapid projected growth. With complex revenue-sharing-based business models, many of these start-up companies promised to launch into "international exponential growth." Few of these start-ups survived. The capital collected was spent, and additional venture financing was no longer available. Yet, for some start-ups, the promised markets opened up. Venture financiers remain optimistic about the prospects of Finnish mobile start-ups and continue to expect good longer-term returns for their risks.

*Hardware*

Finnish telecom companies have been pioneering multiple technology solutions. Finland was one of the first countries to have a fully digitized telecom network, as well as one of the first countries to offer excellent trunk networks for data carriage. In September 2002, Nokia introduced its first 3G phone within Sonera's network and demonstrated its commitment to Finnish know-how. Both wired and wireless networks are among the leading ones in the world. These offer attractive test environments for a variety of mobile solutions.

*Customers*

Finnish mobile technology companies were in a favorable domestic environment during the 1990s. Several players were investing heavily into high-tech innovations and e-business solutions. Mobility was seen as a key high-tech area, and investor interest in mobile companies' activities was high. Multiple trials were carried out for mobile salesforce automation and m-commerce. Furthermore, Finnish consumers were eager to try out mobile applications (such as SMS chatting, SMS dating, TV voting, etc.). By 2002, media companies (e.g., MTV3) started to play stronger roles in m-business. This trend continues to intensify (Karlsson, 2002). These media companies offer their customers interesting content on TV screens, desktop screens, and now also via mobile channels.

## Relationships

Being in the northeast corner of the European Union, Finland lacks the geographic centrality to network its businesspeople into other markets. Only few multinational companies emerged from the country (mainly within the forest and metal industries). Nokia changed this pattern radically. Not only Nokia's own personnel but also its subcontractors and clients have been able to enter into circles much larger than the small Finnish domestic markets. Learning from these operations has been fruitful for the entire Finnish economy.

The mentality of the Finnish business community also changed. More companies started to consider themselves as being a capable and skilled part of the international business community, instead of being a minor actor from "the unknown northern dark country." The domestic networking, often initiated by the national research projects (lead by, e.g., The Finnish Technology Foundation, TEKES), provided excellent opportunities for mutual learning within Finland. Yet, there is still a lot to be learned. Finns are new to service businesses. International markets demand massive and long-term investments from the Finnish companies. Being a domestic league champion offers no guarantee of becoming an influential player in the highly competitive international markets.

## Strategies

During the strongest hype, it seemed that the main focus of mobile companies was to grow at any cost. Capital obtained from investors was spent on expensive international operations and acquisitions. There were strongly held beliefs about "technological pioneering" and about "growing as Nokia did." For a short time, such aggressive expansion strategies boosted the valuations of these companies. Yet, with few exceptions, these commitments became unbearable burdens for the Finnish start-ups—international operations demanded skills and know-how in certain areas, which the companies did not possess.

Only few companies understood the realities of business networks—building sustainable positions in international markets demands much more than sophisticated technological know-how. The largest customer markets (e.g., Germany, France, the United Kingdom, and the United States) are distant from Finland. During 2001 most of the small start-up Finnish companies lost their advantage of being the pioneers of mobile technology.[5] They could not turn the hype surrounding them into profitable customer relationships.

In his research of Finnish mobile start-ups and markets, Leppävuori found the operators' pricing policy to be a significant negative factor impacting the development of the Finnish m-commerce sector. While offering their solutions to Finnish and to various European operators, Finnish content- and application-focused companies faced 80/20 revenue-sharing demands, with operators retaining 80% of the revenue. This is in sharp contrast to Japan, where the revenue sharing is 10/90, with nearly 90% of the revenue flowing back to application developers. It is no surprise that mobile content companies are succeeding in Japan.

The role of the Finnish government in boosting mobile telecom development as well as the crucial role played by Nokia, feature prominently in Leppävuori's findings. Nokia is the largest Finnish multinational company, and the Finnish economy has become dependent on it. When Nokia launches new mobile phone models, such news makes front-page headlines in Finnish media. The immense centrality of Nokia is a boon in booming economic times but poses big risks when Nokia is in a downturn.

## Some Short Case Studies

We now turn to some interesting case studies of small Finnish m-business companies.[6] With nearly 200 digital media start-up firms in Finland, the selected cases (shown as boxed items) only provide an illustrative snapshot. Two of the profiled cases no longer exist, but they are profiled because of their groundbreaking approach to relationship creation for leading IT and entertainment business brand names.[7]

*Figure 4: Factors Influencing the Development of Mobile Services in Finland*

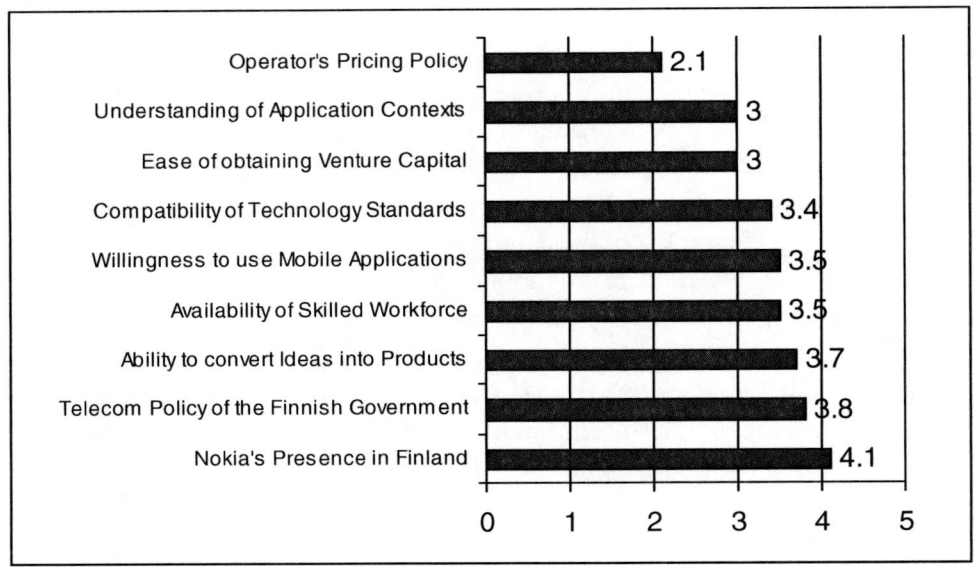

*Source: Adapted and modified from survey of 102 respondents by Leppävuori, 2002*
*5 = Promotes highly; 1= Slows down*

**Riot Entertainment**, *www.riot-e.com*    **RIOT-E**
Year of founding: 1999
(outcome of a merger between three smaller companies)

*Short description of activities:*
Riot-E was created as a leading aggregator of mobile entertainment
and gaming content solutions. Riot-E promised to take leading
entertainment brands into the mobile environment. Riot-E offered
expertise in mobile technology, business, and direct relationships to
mobile operators worldwide, as well as linkages to equipment
manufacturers. Excellent timing of company's market entry and
active business networking brought Riot-E into fruitful and promising
business deals with entertainment brands such as Marvel and New
Line Cinema. Furthermore, Riot-E was flexible and innovative in its
business models with the operators. It was focused on delivering
value-added service for them. Yet, its main challenges were creating

a constant and stable revenue flow. The company obtained nearly 5 million euros as capital investments and expanded its operations quickly. At its peak, Riot-E had over 100 employees and operations in seven countries. Riot-E's business plan was built on an aggressive growth strategy. This high-risk approach led to initial business deal success, but the global ICT crash brought the growth to a standstill.

*Status as of late 2002:*
Riot-E went into bankruptcy after prolonged third-round financing negotiations. Merger with another mobile entertainment company was also considered. While Riot-E did not survive, several of its personnel were able to leverage their accumulated know-how of the global entertainment business into new entrepreneurial activities.

**CodeToys**, *www.codetoys.com*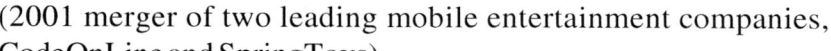
Year of founding: 1998
(2001 merger of two leading mobile entertainment companies, CodeOnLine and SpringToys)

*Short description of activities:*
CodeToys develops mobile entertainment solutions for operators and mobile portals. The company gained worldwide publicity with its mobile versions of the popular "Who Wants to be a Millionaire?" and "Trivial Pursuit" games. Furthermore, the company closed deals with several major entertainment companies, such as Disney and Universal. In addition, CodeToys obtained sound third-round financing from a group of international financiers (Bertelsmann Capital Ventures, AOL Time Warner Ventures, and Motorola) to secure its position in the mobile arena.

*Status as of late 2002:*
CodeToys keeps developing solutions for its customers. The ICT downturn created major challenges. Yet, secured by its financiers, its future still seems promising.

**First Hop**, *www.firsthop.com*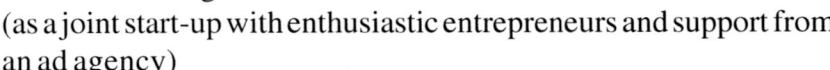
Year of founding: 1997

*Short description of activities:*
First Hop started its operations as a subcontractor for Web design agencies. The company rapidly developed its competencies, especially in Java-based solutions. First Hop created its own software products for online and mobile publishing and content management platforms. First Hop obtained venture financing and had international customers for its solutions. It has been judged to be one of the most promising mobile developing companies offering solutions to telecom operators and directly to corporations. Yet, like other Finnish start-ups, it has faced growth challenges and major profitability pressures from its financiers.

*Status as of late 2002:*
First Hop continues to operate and develop its solutions for its customers. The latest expansion has been to start providing solutions to Southeast Asia.

**Sulake Labs**, *www.sulake.com*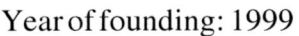
Year of founding: 1999
(as a joint start-up with enthusiastic entrepreneurs and support from an ad agency)

*Short description of activities:*
Springing from the hobby of its multimedia-enthusiastic founders, Sulake Labs created one of the first truly profitable "virtual goodie" environments. Their concept "Habbohotel" is a virtual 3D community in which registered members can interact, move, chat, play games, dance, etc. Furthermore, users can create their own rooms in which they can purchase virtual furniture and other decorations with their mobile phones. Surprisingly, the concept has proven to be very successful, and Habbohotel has several thousand active members. The original Finnish operation, "Hotel Kultakala," is offered in

cooperation with a Finnish telecom operator and has become profitable for both parties. In addition, the solution won several awards in international digital design contests.

*Status as of late 2002:*
Sulake continues to develop its technology solutions. It opened its Fuse software solution for developers worldwide. Sulake also expanded into international markets, with operations in the United Kingdom and elsewhere.

**WapIt**, *www.wapit.com*
Year of founding: 1999

*Short description of activities:*
WapIt is often seen as the pioneer of the Finnish mobile start-ups. It was among the first start-ups to obtain millions of venture capital for its international growth. It gained high publicity (e.g., in *Business Week*, *Economist*, and *Financial Times*) as being one of the first global mobile solution companies. It pioneered in revenue share-based business models with the operators. It was among the first companies to expand its operations with partnership deals with Hewlett-Packard and Nokia. Finally, however, WapIt also pioneered as one of the first mobile companies to crash. The company tied itself into unfinished technology and became a victim of the overall troubles of WAP. Its business plans were based on exponential consumer adoption of mobile solutions. The rosy expectations, strong publicity, and technology forecasts allowed WapIt to attract investors (such as Durlacher) worldwide. Yet, the planned business never materialized. Though the company had innovative solutions and technology, it had to discontinue its operations in 2001.

*Status as of late 2002:*
Having run out of additional financing, WapIt filed bankruptcy in 2001.

**Iobox**, *www.iobox.com*
Year of founding: 1999
(as a start-up)

*Short description of activities:*
Iobox was one of the first Finnish companies to start offering free Web-based e-mail to its registered users (as GNW mail service). In 1999, it changed its name into IoBox and started offering mobile e-mail and calendar access, as well as ringtones and operator logos. The company obtained venture capital and expanded rapidly into international markets. In 2000, the company was acquired by Spanish telecommunications operator Terra Movile at the high price of nearly 160 million euros. Promised expansion turned into rapid downgrading in 2001. Subsequently, Iobox discontinued its Finnish operations in late 2001 and focused solely on major European markets. Iobox was also integrated more tightly into its Spanish parent's operations.

*Status as of late 2002:*
Iobox continues as part of Terra. It operates currently in four countries (the United Kingdom, Germany, Spain, Brazil). It still has a few Finnish employees. The company, however, has totally withdrawn from Nordic operations.

**Jippii**, *www.jippii.com*
Year of founding: 1997
(through a merger of Finnish ISPs—its initial name was Saunalahden Serveri)

*Short description of activities:*
Jippii was created originally as a challenger in the rapidly developing Finnish Internet service provision markets. During the hype, Jippii also entered the mobility markets. The company was among the first portals and virtual mobile operators to start offering "goodies" like mobile phone ringtones and operator logos to its users. Rapidly, this virtual "goodie" business brought Jippii from high losses into nearly profitable business. In its aggressive expansion into international markets, Jippii entered into a German local access provisioning

business. This operation became nearly fatal for the company. Jippii had to withdraw from Germany at a very high cost and obtain rescue financing from its main investors.

*Status as of late 2002:*
Jippii recently stabilized its position among leading Finnish portals. It is still offering innovative mobile solutions for its customers. Yet, it is more focused on core operations—Internet access provisioning, especially offering broadband solutions. Furthermore, the company runs its operations domestically again, with the original Saunalahti-brand name.

**Sonera Zed**, *www.zed.com*
Year of founding: 1999

*Short description of activities:*
Springing from its domestic mobile phone service operation, Sonera formed its internationally focused mobile portal in 1998. Zed was one of the first mobile portals and immediately created high expectations throughout the business community. The company was estimated in early 2000 to be worth billions of euros if publicly listed. Sonera invested tremendous amounts of money into Zed's expansion into European markets and brand creation. Yet, high investments brought back only marginal revenue. The consumers did not buy into Zed's WAP and SMS-based offerings. Zed was making money with operator logos and ringtones, but not enough to cover the investments put into it. In early 2001, Sonera planned to list the company, but the plans were postponed due to the downturn. Zed withdrew several of its plans and went into massive restructuring of its operations.

*Status as of late 2002:*
Yahoo! Mobile acquired 15% share of Zed from Sonera in 2002. Yahoo also has an option to acquire the rest of the company. Currently, the two companie*s are planning marketing cooperation and possible joint development projects.*

## Some Lessons from Finnish Pioneering Cases

The mobile sector is very much in flux. The early cases described here, however, yield some important lessons for m-business success:

- Simple, entertaining items (games, ringtones, logos) have significant revenue potential if volume can be built up and sustained.
- Business network relationships, especially with deep-pocketed operators, offer a substantial degree of financial flexibility and thereby the ability to reorient the business models. In terms of Figures 2 and 3, the proximity and strength of the relationship between a startup and the telecom network operator seems to improve survival chances and prospects for financial viability.
- Business models based on unproved technologies and on assumptions of enduring exponential growth are extremely risky.

In the sections to follow, we provide the conceptual elements that can help in explaining the Finnish pioneering cases as well as in assisting future m-business strategies the world over.


# A GENERIC BUSINESS NETWORK MODEL FOR M-COMMERCE

## Business Network Theory

Researchers attempting to understand the developments in the mobility-related markets suggest utilizing business network theories as the appropriate descriptive and analytical approach to grasp this sector (Durlacher, 2001; Leppävuori, 2002). These theories originate from the groundbreaking work of the Scandinavian-English IMP-research group (Håkansson & Johansson, 1992; see also Håkanson & Snehota, 1994; Johansson & Mattson, 1988; Axelson & Johansson, 1992). Originally, the researchers working with business network theories analyzed dyadic relationships between industrial buyers and sellers. More recently, the approach was widely accepted to describe markets as *value-webs*, networks of interconnected actors, each possessing limited amounts of resources and each performing specific market activities. Activities are performed to obtain as dominant a market position as possible.

*Table 3: Mobile Telecom Business Activity Set*

| Core Activities | Supporting Activities | Infrastructure-Related Activities | Research and Development Projects | Business Network-Building Activities | Creation, Enforcement of Market Rules |
|---|---|---|---|---|---|
| - Access provision<br>- Data delivery and transport<br>- Service provision<br>- Service hosting | - Advertising (traditional and digital media)<br>- Consulting<br>- Training<br>- Content creation<br>- Content aggregation | - Equipment manufacturing<br>- Software production<br>- Network building and maintenance | - Own R&D projects<br>- Joint R&D projects | - Formal and informal negotiations on financing, project cooperation<br>- Subcontracting agreements<br>- Research and development | - Regulations<br>- Laws<br>- Standards |

*Source: Modified and Adapted from Pelkonen, Pohto and Wirén, 2001b.*

## Business Network View of Mobile Telecom Markets: Activities and Resources

Mobile telecom markets can be viewed as business networks, wherein a variety of actors (operators, governments, infrastructure providers, device makers, value adders, users, and so on) possess resources, perform activities, and are in relationships that are established or evolving. The resultant interactions influence the market positions of the actors. To begin analyzing mobile telecommunications from a business network perspective, we first look at the market activities (Table 3). These are the operations that companies, organizations, and individuals carry out in the market. Mobile telecommunication business activities can be categorized into main activities and supporting activities. At the core of the telecom business (fixed or mobile) are *access provision, data delivery, service provision*, and *service-hosting activities*. These are supported by multiple other activities, such as advertising, training, and content creation. Furthermore, in the telecom industry, multiple types of equipment are required. Thus, *equipment manufacturing and maintenance* form one activity area. Companies invest in the development of new solutions for their customers, and this *research and development* is carried out by an actor's own or joint efforts. In addition, companies constantly aim to strengthen their strategic positions by using a variety of *business network-building activities*. These entail formal and informal negotiations, different ways of subcontracting and outsourcing, and joint research and development. Finally, various governmental and industry organizations also aim to influence and standardize the competition terms in the markets by implementing *market rule creation and enforcement activities.* In Table 3, the business activity set of the mobile telecommunications sector is summarized.

Mobile telecom business activities are based on various resource pools. These can be categorized into human, software, hardware, organizational, and financial resources (see, e.g., Holmlund & Kock, 1995). The key resources for operation are the company's employees. Also, telecom operations are capital intensive and require substantial investments into switching equipment and transmission networks. Thus, these form the second key resource group. The telecom business resource categorization is presented in some detail in Table 4.

## Business Network View of Mobile Telecom Markets: Actors and Relations

There are a variety of mobile telecommunications market actors. The core actors are the *telecom network operators.*[8] The operators possess the key understanding of the telecom service users: consumers and corporate customers. Through their business development efforts, the operators generate revenue opportunities from the users. In fact, the end users represent the only sustainable revenue source for all market actors. The money exchanged over usage of telecom services originates from the end users and is distributed among the other actors in the marketspace. Almost without exception, the network operator holds a dominant or at least controlling position in this crucial monetary flow.

*Table 4: Mobile Telecom Business Resource Set*

| Human | Hardware | Software | Organizational | Financial |
|---|---|---|---|---|
| • Management<br>• Technical<br>• Design<br>• Maintenance<br>• Service<br>• Sales<br>• Marketing<br>• Other | • Office premises<br>• Production machinery<br>• Personal computers and servers<br>• Network equipment<br>• Cables and control equipment<br>• Locations for network equipment | • Licenses<br>• Intellectual property rights<br>• Proprietary contents<br>• Production process knowledge<br>• Knowledge about technology<br>• Knowledge about customers<br>• Production software | • Strategies<br>• Goals<br>• Organizational culture<br>• Organizational structures | • Finance for operations (e.g., R&D, commercialization, internationalization)<br>• Capital valuation in the stock market |

*Source: Modified and Adapted from Pelkonen, Pohto and Wirén, 2001b.*

Also connected directly to the end users are various *mobile media (portals)*. These are business activities of the network operator or of an external party. If an external actor is the portal provider, the activities of such an external actor nearly always remain dependent on the business relationships between network operators. End users are, in most cases, charged only via the mobile operator's billing systems, regardless of the type of third-party or value-added service usage. At their inception in 2000, mobile portals attracted major interest in the business community as new media, but their success so far has been limited.[9]

For a mobile portal or a network-delivered service to be useful, interesting content needs to be created. For this, various content-related actors are needed. Content value creation can be described as a five-step process (see Figure 5).[10] Mobile telecommunications follows this typical content creation chain, and each step can have its specialized actors. A mobile portal or an operator most often carries out content marketing. The operator nearly always handles the distribution of content.

It is technology-oriented companies that have mainly formed the mobile industry. These could fall into two main categories: *service and application enablers* and *enabling technology providers.* The former consists of companies that develop and market their solutions (normally software products) to mobile operators or mobile media, occasionally also directly to corporate end users and even to consumer markets. The latter can be further categorized into four main groups: developer tool vendors; service platform and component creators; network infrastructure vendors; and device manufacturers. These companies sell their products mainly to telecommunications operator globally. Yet, they also have direct customer relationships to service enablers and mobile mediums. In addition, mobile device manufacturers sell their products to end users through their own distribution channels.

The m-commerce business field involves several connected actor groups. *Financiers* seek profitable investment targets. In the period of upturn in the 1990s, large risks were taken by seeking companies with strong chances for success and then investing in such firms. Only few of these investments proved viable. A large number of ventures failed to fulfill their business plans and ran

*Figure 5: Generic Content Creation Process*

*Figure 6: Mobile Telecommunications Business Field*

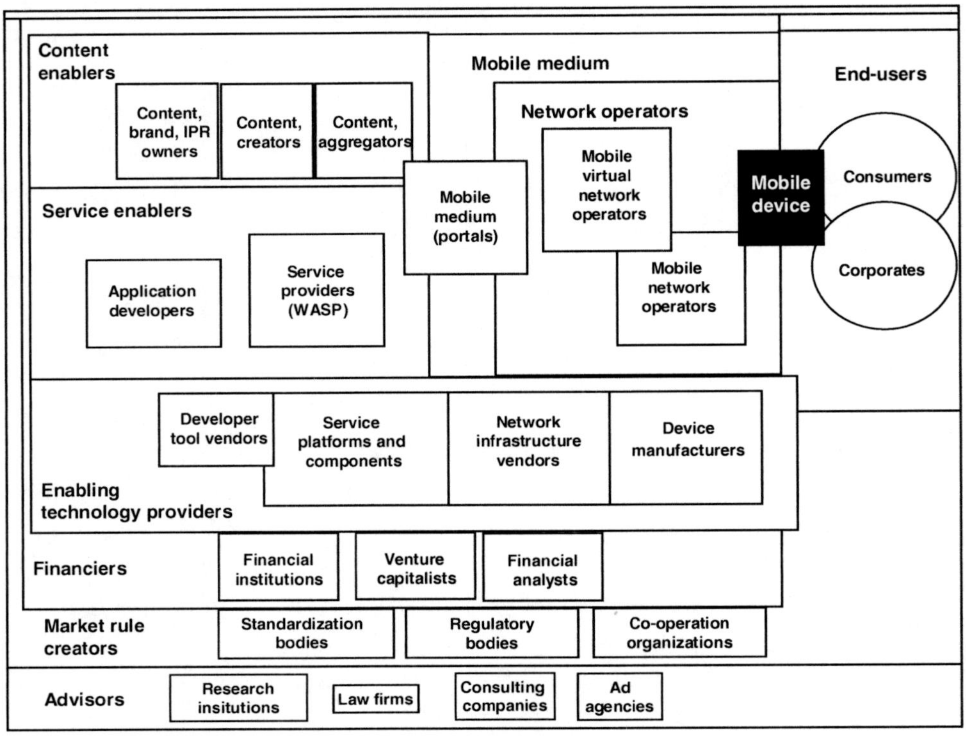

short of money. This is bound to impact the industry's prospects, at least in the short term. *Regulators* and standardization organizations work as the market rule creators for companies in the m-business fields. Since the early 1990s, their impact has been important to the whole industry due to increasing global telecom deregulation. Finally, as in any business, a large number of *advisory service companies* also operate in the mobile communications field. These include legal, marketing, and business advisors, as well as various research institutions.

The business field of mobile communications could be illustrated as an actor web (see Figure 6). On the top of the figure is the content creation value chain, while below it are the related actor groups.

By combining the business network elements of Actors, Resources, and Activities, an overall business network model for mobile telecom can be formed (see Figure 7). Telecom actors compete and cooperate to obtain the most valuable, skilled, and scarce resource pool to perform their activities. Very

*Figure 7: Mobile Telecommunications Business Network*

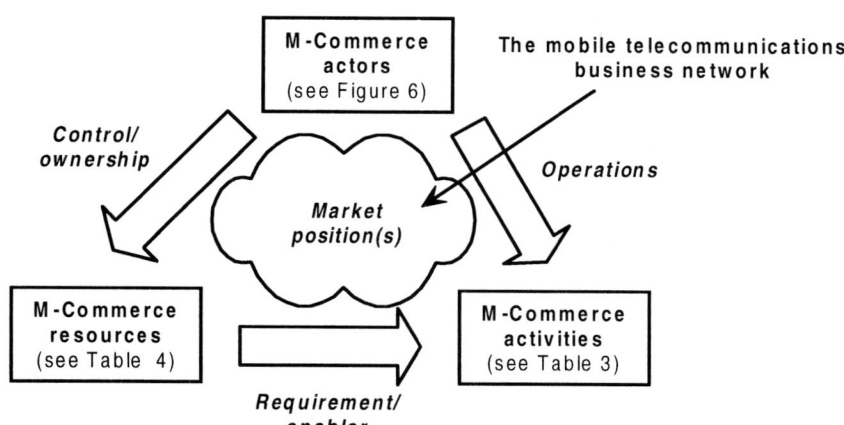

often, due to their history and control over the end users, telecom operators hold a dominant position in the markets. The second powerful market group is comprised of the equipment manufacturers. By 2002, media companies increased their importance in the m-business fields. A common factor characterizing these three groups is the close relationship to end-users. From the perspective of start-up firms, the m-commerce field is challenging. Business success is highly dependent on the firm's abilities to create a compelling competitive position, attract a large customer base, and generate sustainable revenue flows.

For mobile application developers and content-oriented start-ups, a lot of partnering is needed for successful business operations. As mentioned earlier, ultimate business models have to rely on increased revenues from the end users. In practice, this means that the network operator has to obtain a higher ARPU (Average Revenue per User) by offering its customers (the end users) services that they are willing to pay for, at levels over and above the charge for their conventional mobile voice and data usage. It is this incremental revenue that the operator is able to share with application developers, technology enablers, and content enablers. Such incremental revenue has so far proven to be elusive, at least in the European markets. Increments in revenue through offering these value-added services have been much smaller than expected. Therefore, the business plans based on revenue share-based earnings models for start-ups have very often failed.

# CONCLUSION AND FUTURE DIRECTIONS
## Understanding Position and the Resource Pool

In the mobile business, positioning oneself in the competitive landscape is of utmost importance. Increasing pressures for revenue generation characterize the current, emergent phase of this industry. For all industry actors, this leads to seeking more sales or more efficiency through their own efforts and via joint activities. This calls for implementing relationship "orchestration," based on internal and potential partner's resources, and seeking competitive advantages via efficient resource utilization as well as creative resource blending.

The main challenge for several pioneering mobile Finnish companies was to obtain and allocate their resource pools to match ambitious growth strategies. High risks were taken in business expansion, often beyond the limits of the firm's control. A few firms benefited from being among the first movers within the mobile arena, but for many, these risks became overwhelming. In this dynamic and emergent period, the mobile industry keeps transforming, quarter to quarter; only time will tell which of the Finnish m-business companies will survive in the long term.

## Identification of Crucial Partnerships and Activities

Planning and implementation of network-based business operations can be challenging. The mobile business-network framework presented in this chapter is geared to help companies and researchers identify and analyze the market positions of key actors. The business-network diagramming and assessments can be used to map the firm's own position and the positions of external actors, resource profiles, business activities, and partnering actions. The partner search should not be confined to the traditional geographical limits of the company. Instead, strategic match seeking should be extended to international markets. This is especially necessary for m-business companies based in small economies, such as in Finland.

In most global markets, the mobile telecom operator holds the pivotal position for revenue generation. In some countries, the dominant actor can also arise from other parts of the value-web, such as device and equipment maker Nokia in Finland. In the converging ICT landscape, important strategic positions can also arise from surprising areas, such as the success of TV channels in SMS+TV+Chat solutions.[11] The telecom operators, however, have the most entrenched customer base, functioning billing systems, and understanding of consumers' mobile phone usage patterns. In addition, the operators operate their own or leased mobile networks. For mobile start-up success, creating smooth-functioning relationships with operators is undoubtedly a key requirement. If a start-up cannot provide sustainable revenue generation streams from the operators' end users, then it cannot be viable in the long run.

An alternative to sustained operator relationships is to have a short-term business focus with aggressive marketing and brand creation. During the ICT hype of the 1990s, these kinds of operations were sufficient for start-ups. In the future, the situation will be far more challenging. M-commerce firms need to show recurring returns on investment, not only short-term revenue potential but also sustainable long-term revenue streams.

## Mobile Earnings Logics

Multiple revenue-sharing-based business models were already tested across global mobile markets. Yet, pioneering experience from Finland showed that the trend is from complex business models to simpler ones. Advertising, subscription fees, and transaction commissions remained the main revenue generators. Early experience indicates that business models based on shared micropayments (sharing the end users' fractional euro/dollar payments for ring tones or icons) are challenging to administer and sustain. As content aggregators, mobile media (portals) benefited from such micropayment-based revenue, but the revenue passed backward to individual content creators or application vendors has been marginal. The start-ups were given only minimum shares of the possible revenue from mobile innovations that these start-ups created and made available to the portals or operators.

As mentioned frequently in this chapter, controlling customer information is critical for market position improvement. The actors that can actually send the bill to the end user—mostly telecom operators—hold the key role in the mobile value-webs. Regardless of technological development, this reality will not change. Mobile start-ups sometimes aim to bypass the telecom operators by offering their own access gateways and service portals. Yet, obtaining a critical and sustainable mass of paying customers for such specialized gateways is not easy. Telecom operators, regardless of their current downturn, have deep pockets to fund the high investment costs required for m-commerce operations. Start-ups, especially after the dot.com crash, are struggling with their survival and are thus in very weak competitive positions. Telecom operators may end up acquiring innovative, but resource-starved start-ups at bargain basement prices not reflective of the long-term value of such start-ups.

By 2000, throughout the mobile industry landscape, actors started facing harsh economic realities. The time for hype was over; and the mobile industry was becoming a business similar to any other business. Free lunches—in the form of fast and loose venture capital—ceased to exist, and business models with long-term viability became keys to success. The Finnish experiences outlined in this chapter, along with concepts from business network theories, should help m-commerce players make their business models viable and robust.

# ENDNOTES

1    In its early years, like most other telecom firms worldwide, Telecom Finland (TF) operated as a division of the Finnish national post office. TF changed its name to Sonera in 1997. Now, Sonera and Sweden's Telia merged.

2    The U.S.-based author of this chapter had an interesting experience in Helsinki in 2000. From his apartment in Helsinki, the author called the mobile phone of a Nokia executive. The Nokia executive said: "I was surprised to get this call. I hardly ever get a call from a fixed line!"

3    ICT refers to Information and Communications Technologies.

4    See, e.g., www.medialinnakkeet.com for more information.

5    By 2002, Japan and South Korea were being mentioned as the most advanced mobile countries globally.

6    The case studies are based on publicly available news materials from www.digitoday.fi, www.talentum.com, and www.itviikko.fi.

7    A comprehensive listing of Finnish companies operating in the ICT industry can be found at, e.g., http://www.swbusiness.fi/.

8    In mobile telecommunications, there are two kind of network operators: a mobile network operator (MNO, incumbent operator) builds and operates its own network, while a mobile virtual network operator (MVNO) rents transmission capacity from the MNOs but operates under a separate brand and with separate customer billing systems.

9    See, e.g., http://www.gsacom.com/, Mobile Portal Surveys.

10    Such a model can be applied to any content creation business, be it telecommunications or another sector.

11    See, e.g., http://www.franticmedia.com/, http://www.waterwar.tv/, or www.matchem.com.

# REFERENCES

Eqvitec Partners & Durlacher Ltd. (2001). UMTS Report: An investment perspective. Durlacher Research Ltd., London.

GSM-Association. (2001, ongoing). Mobile Portal Surveys. Available at: http://www.gsacom.com.

Håkansson, H. & Johanson, J. (1992). A model of industrial networks. In *Industrial networks: A New View of Reality* (pp. 28-34). London: Routledge.

Håkansson, H. & Snehota, I. (1992). Analysing business relationships. In *Developing Relationships in Business Networks* (pp. 24-49). London: Routledge.

Holmlund, M. & Kock, S. (1995). Buyer perceived service quality in industrial networks. *Industrial Marketing Management, 24*, 109-121.

Johanson, J. & Mattsson, L. -G. (1988). Internationalisation in industrial systems—A network approach. In *Strategies in Global Competition* (pp. 287-314). New York: Croom Helm.

Karlsson, J. (2002). Mobile earnings logics. Lecture at Mercuria Business School. October 2. Vantaa.

Kuokkanen, N., Toivola, T., & Väänänen, T. (1999). Uusmediatoimiala Suomessa 1999. LTT-Tutkimus Oy. Helsinki. www.uiah.fi/mediastudio/tutkimus.html.

Leppävuori, I. (2002). Analysis of the Finnish mobile cluster—Any potential in mobile services? Ministry of Transport and Communications. Helsinki, Finland. Available at: http://www.mintc.fi/www/sivut/dokumentit/julkaisu/julkaisusarja/2002/a282002.htm.

Meristö, T., Leppimäki, S., & Tammi, M. (2001). ICT-osaaminen 2010—Tietoteollisuuden ja digitaalisen viestinnän osaamisen ennakointi, Åbo Akademi/Institute for Advanced Management Systems Research, CoFi Report 1/2002. Turku.

Pelkonen, T. (1999). *Resource-based internationalization of professional business services—A case study of the Finnish new media industry.* Helsinki School of Economics, Master's Thesis. Helsinki, Finland.

Pelkonen, T., Pohto, P., & Wirén, L. (2001). Uusmedia aikuistumisen kynnyksellä. Uusmedian osaamiskeskus ja LTT-Tutkimus Oy. Helsinki. Available at: http://www.uiah.fi/koulutuskeskus/uusmedia.pdf.

**Chapter VI**

# Understanding the Mobile Consumer

Constantinos Coursaris, McMaster University, Canada

Khaled Hassanein, McMaster University, Canada

Milena Head, McMaster University, Canada

## ABSTRACT

*The recent surge of interest in mobile commerce (m-commerce) is fueled by consumer interest in being able to access business services or to communicate with other consumers anytime and anywhere. It is also motivated by the interest of the business community to extend their reach to customers at all times and all places. Businesses that aspire to succeed in this market must have a deep understanding of the interests and concerns of the mobile consumers in using wireless applications. With this in mind, this chapter provides an analysis of this emerging market from a consumer's perspective. A consumer-centric m-commerce model outlining the various wireless interaction modes of the mobile consumer (m-consumer) is presented, followed by a discussion of the needs and concerns of the m-consumer. An m-commerce value network is then presented, outlining the roles of the different players within this industry. The various business applications developed to address m-consumer needs are then presented. Finally, a global m-commerce market overview is provided, and some future trends are outlined.*

# INTRODUCTION

From its inception and over the last two decades, the Internet has undergone significant change. It dramatically adapted to suit new network technologies and telecommunication services. The subsequent ability to engage in transactions for personal or professional use over the Internet has come to be known as electronic commerce or e-commerce. The most recent trend of e-commerce involves expanding the services offered and extending the reach to customers through powerful affordable computing and communications in portable forms (i.e., laptop computers, two-way pagers, PDAs, cellular phones). The mobility associated with these devices has resulted in naming this new trend mobile commerce or m-commerce (Leiner et al., 2002). This trend is fueled by a consumer interest in being able to access business services or to communicate with other consumers anytime and anywhere. It is also motivated by the interest of the business community to extend their reach to customers at all times and at all places.

This chapter starts by exploring the similarities and differences between m-commerce and e-commerce. In particular, contrasts will be made in the areas of communication modes, Internet access devices, development languages/communication protocols, and enabling technologies. A consumer-centric m-commerce model outlining the various wireless interaction modes of the mobile consumer (m-consumer) is then presented, followed by a discussion of the needs and concerns of the m-consumer. M-consumer needs are then matched with relevant concerns, and special emphasis is given to the important areas of cost, wireless security, and wireless privacy. An m-commerce value network is then presented, outlining the roles of the different players within this industry. The roles of the various players within the m-commerce value network in addressing the m-consumer concerns are then discussed. The various business applications developed to address m-consumer needs are then presented and classified according to the different need areas. We also summarize the current technologies in support of such applications, indicating any shortcomings of such technologies that might stifle the consumer adoption rates of particular wireless business applications. Future technologies that could resolve such issues are also outlined. A global m-commerce market overview is provided. The chapter ends with a discussion and some future trends.

# M-COMMERCE OVERVIEW

The name "m-commerce" arises from the mobile nature of the wireless environment that supports mobile electronic transactions. Devices, including digital cellular phones, personal digital assistants (PDAs), pagers, notebooks, and even automobiles, can already access the Internet wirelessly and utilize its various capabilities, such as e-mail and Web browsing (Little, 2001). Because m-commerce encompasses any electronic activity that utilizes wireless technology, it may be viewed as a subset of e-commerce. Furthermore, m-commerce and e-commerce share fundamental business principles, but m-commerce acts as another channel through which value can be added to e-commerce processes. It also provides for new ways through which evolving customer needs could potentially be met.

## Contrasting M-Commerce and E-Commerce

The m-commerce and the e-commerce business environments and activities have much in common. This is the case, because they involve much of the same functionality in terms of facilitating electronic commerce over the Internet. However, some differences exist in the mode of communication, the types of Internet-access devices, the development languages and communication protocols, as well as the enabling technologies used to support each environment. Differences in these four areas are explored below in more detail (Little, 2001).

- **Communication mode:** E-commerce is mainly conducted through a wired connection, while m-commerce mainly operates over wireless networks. This is a fundamental difference between the two environments, as m-commerce overcomes the barrier of a fixed location, and customers can engage in commercial activities anytime, anywhere using various forms of wireless communication devices.
- **Internet access devices:** While e-commerce is conducted mainly through wired desktop and laptop computers, m-commerce is conducted

through wireless devices (e.g., cell phones, PDAs, wireless-enabled laptops). Because wireless devices tend to be used by a single user who carries the device at most times (and because the location of the device can be tracked), there is enhanced opportunity to offer personalized products/services, albeit privacy concerns are escalated because of this tracking/personalizing ability.

- **Development languages and communication protocols:** Hypertext Markup Language (HTML) runs the Web on wired networks, whereas on wireless networks, wireless devices are running on one of two variations of HTML: Wireless Markup Language (WML) or compact HTML (cHTML). The need for WML and cHTML is due to mobile devices having to comply with new communication protocols [e.g., the Wireless Application Protocol (WAP) and i-mode®]. Different from the wired Web's Hypertext Transfer Protocol (HTTP), these new protocols present issues of compatibility and functional limitations.
- **Enabling technologies:** Several of the existing technologies that enable e-commerce with relative ease (e.g., persistent cookies, i.e., cookies that are stored between user sessions) are not currently supported over wireless networks and cannot be utilized by m-commerce.

## M-Commerce Technology

As indicated in the previous section, the greatest difference between m-commerce and e-commerce lies with what and how technology is being used. In this section, three areas of technology that are fundamental for m-commerce will be examined in further detail: wireless networks, wireless protocols, and wireless devices.

### Wireless Networks

Wireless networks provide the backbone of m-commerce activities. Users can transmit data over these networks between mobile and other computing devices through the use of wireless adapters without requiring a wired connection. The first wireless networks were introduced as early as 1946, but a major milestone was the introduction of the Advanced Mobile Phone System (AMPS) that marked the arrival of cellular systems in 1983 in the United States. The AMPS is an analog system used for voice communication (3GAmericas, 2002). AMPSs represented the first generation of cellular systems (hence, it is commonly referred to as "1G").

*Table 1: Wireless Network Technologies: Current and Future[4]*

| Region | Current Network (2/2.5G) | Future Network (2.5/3G) |
|--------|--------------------------|-------------------------|
| US | TDMA, D-AMPS, CDMA, GSM, Mobitex, CDPD | CDMA2000 (2003) |
| Europe | Mobitex, GSM, HSCSD, GPRS | EDGE, W-CDMA (2002) |
| Japan | cdmaOne, PDC, W-CDMA | W-CDMA, cdmaOne (2002) |

The evolution of wireless networks continued with the implementation of second-generation (2G) systems that were introduced in the 1990s. Several of these systems (e.g., TDMA, CDMA, GSM)[1] were also used primarily for voice applications, with the exception of the Short Message Service (SMS) capability offered by the GSM network. A recent upgrade of the 2G networks is referred to as 2.5G wireless networks (e.g., HSCSD, GPRS, EDGE)[2]. Being either circuit-switched or packet-switched, these networks are primarily intended to allow for increases in data transmission rates and, in the case of packet-switched networks, an "always-on" connection (Peck, 2001).

The hype surrounding wireless networks, however, revolves around the third-generation (3G) systems, expected to be deployed over the next few years, with certain regions already having access to them (e.g., Japan). These networks are commonly referred to as IMT-2000 on a global scale, and regional implementations are uniquely named (e.g., CDMA2000 in North America, W-CDMA/UMTS in Europe and Japan, cdmaOne in Japan).[3] Along with voice functionality, 3G networks support higher-speed transmissions for high-quality audio and video enabled through high-bandwidth data transfers, as well as provide a global "always-on" roaming capability (Peck, 2001).

Shown in Table 1 are wireless network technologies that are currently in use or are expected to be rolled out in the regions of North America, Europe, and Japan (Peck, 2001). In Figure 1, the path to the anticipated ubiquitous 3G environment is illustrated (ITU, 2001).

*Wireless Protocols*

While wireless networks evolved, the two main communication protocols, WAP and i-mode, experienced their own evolutions. Phone.com, Ericsson, Motorola, and Nokia introduced WAP in 1997. WAP progressed from enabling basic functionality, such as WML and WMLScript communications,

*Figure 1: Evolution of Wireless Networks (Adapted from ITU, 2001)*

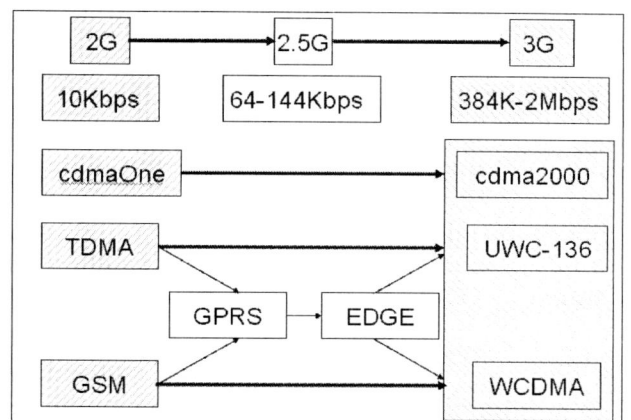

in its first release, to supporting graphics, voice-enabled actions (i.e., wireless Web browsing), and video, as announced in the release of WAP 2.0, at the end of July 2001 (WAP Forum, 2001). On the other hand, i-mode was introduced in 1999 by NTT DoCoMo and has grown in popularity to support 30 million users in less than three years. The capabilities of i-mode were enhanced during 2001 through the introduction of i-appli, which incorporates JAVA and Secure Socket Layer (SSL) encryption capabilities; i-area, which provides location-specific information such as weather, local guide, maps; and traffic; and i-motion, which enables the viewing of video clips (NTT, 2002).

*Wireless Devices*

Until recently, wireless devices could be placed in three distinct categories: wireless phones, wireless PDAs, and wireless laptops. Recently, however, hybrid products have been introduced that combine features from two or all three categories with the intent of providing enhanced capabilities to mobile users.

Mobile phones have been around the longest and have experienced the greatest changes since their inception. In the beginning, analog cellular phones were used exclusively for voice communications; next, digital phones were introduced, initially for voice communications but with added features (e.g., Call Display) and were later further enhanced with additional capabilities (e.g., Instant Messaging).

PDAs experienced their own evolution, beginning as organizers for personal information with limited functions (e.g., "To Do" lists, calendar). Currently, some PDAs have wireless transmission and Web-browsing capabilities. The major operating systems for PDAs are Palm OS (e.g., Palm, IBM WorkPad PC, Handspring Visor), EPOC (e.g., Ericsson R380, Nokia 9210 Communicator, Psion), and Windows CE (Compaq iPac, HP Jornada, Casio E-125).

Wireless laptops include notebooks or portable PC browser clients that are wirelessly Web-enabled (e.g., IBM ThinkPad T20 connected with a GSM mobile phone through the infrared port). Although these devices are capable of supporting m-commerce activities, they do not represent the main point of access for such activities due to their relatively larger sizes and heavier weights compared to other mobile devices (Peck, 2001).

The most recent development in mobile devices was the introduction of "smart phones." These are mobile devices capable of tasks ranging from e-mail retrieval now to video and music streaming in the near future. Smart phones are a combination of cell phones and PDAs (e.g., Kyocera QCP™6035 Smart Phone, Samsung SPH - I300) (Pocket, 2001).

# M-COMMERCE CONSUMER INTERACTIONS

To better understand the value proposition that m-commerce presents to consumers, it is important to identify the m-consumer interaction modes within a wireless environment. By reflecting on the m-consumer's possible activities, one could identify the following entities with which interaction may be required or desired to various degrees:

- **Businesses:** Involving a Wireless Business-to-Consumer ($W_{B2C}$) interaction mode. It is important to note that most such interactions would naturally involve a Wireless Consumer-to-Business ($W_{C2B}$) interaction mode as well.
- **Consumers:** Involving a Wireless Consumer-to-Consumer ($W_{C2C}$) mode of interaction.
- **Personal networks:** Involving a Wireless Consumer-to-Self ($W_C^2$) interaction mode.

These entities and interaction modes are illustrated in Figure 2, where the entities are shown in rectangular boxes, examples of these entities are shown

*Figure 2: A Consumer-Centric M-Commerce Model (Coursaris &
Hassanein, 2002)*

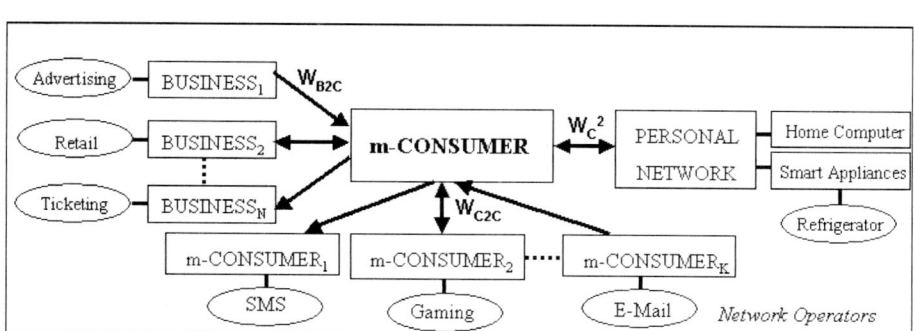

in ellipses, and the arrows indicate the direction of the interaction (i.e., who
initiated the action). Some interactions, shown by single-directed arrows, are
performed solely by one entity and do not necessarily receive a response by
other entities (e.g., stock alerts). Other interactions, shown by double-directed
arrows, require the active involvement of both parties in a mobile transaction
(e.g., mobile retailing or m-tailing). Any of the above types of mobile transactions require at least one entity to be using the wireless channel and may involve
some wired participants. Finally, network operators are included in this model
but are not linked, because they provide the necessary infrastructure for these
relationships to take place or act as facilitators for supporting m-commerce-
related activities.

   **Businesses** refer to individuals or organizations that a consumer may need
or want to interact with wirelessly for business-related purposes. In addition,
consumers may be at the receiving end of an interaction initiated by businesses.
For the purposes of this chapter, $\mathbf{W}_{B2C}$ is used to refer to this type of interaction,
without paying attention to which party initiated the interaction. Some examples
of business applications in this area include retail and advertising offers directed
at m-consumers. These applications are made available through the combined
efforts of all members (excluding customers) of the value network to be
introduced in the section "Addressing m-consumer concerns" (see Figure 3).

   **M-consumer** refers to individuals that a consumer may need or want to
interact with wirelessly for personal purposes. Some examples of this $\mathbf{W}_{C2C}$
type of interaction include communications (e.g., SMS or e-mail) and entertainment (e.g., gaming in a multiplayer format).

   **Personal network** refers to the server that a consumer owns and may
want to access wirelessly for personal purposes. This type of interaction is

identified through the notation $\mathbf{W}_C{}^2$. Examples of this type of interaction include a mobile user engaged in wireless communications with his or her home computer and its network, as well as any smart appliances that may be connected to that network (e.g., a refrigerator).

# M-CONSUMER NEEDS

Five primary needs can be identified that yield demand for m-commerce services. These five needs stem from the mobility associated with the enabling devices, so the context for each of them revolves around the theme of "anytime, anywhere" accessibility. These needs and the subsequent discussion may also apply to business customers to some degree, but specifics associated with the business segment are not explored in this chapter, as the focus is on the end consumer. These needs are as follows:

- **Connectivity needs:** Connectivity provides the basic platform on which wireless communications take place. In an ubiquitous wireless environment that overcomes geographic (i.e., location of the consumer) and compatibility (i.e., interoperability of networks) constraints, consumers become capable of true "anytime, anywhere" communications.
- **Communication needs:** M-consumers communicate with others for business or personal purposes (i.e., with other consumers or personal networks), and their communications may be carried out within information, entertainment, and commerce contexts.
- **Information needs:** M-consumers need access to information that may be static (e.g., Yellow Pages-type directory) or dynamic (e.g., cross-referencing of wireless Web sites for prices or specifications of a particular product). In addition, consumers may be interested in location-specific information (e.g., finding a restaurant based on the user's search criteria and current location).
- **Entertainment needs:** Wireless devices can provide users with practical entertainment solutions, such as access to games or leisure information.
- **Commerce needs:** Two main elements are required to enable m-consumers to conduct m-commerce transactions: presentation of product/service information and a wireless payment mechanism. The value in consumers making payments wirelessly arises from the convenience it

offers. For example, mobile users might not require coins/bills to make certain physical purchases (e.g., from vending machines), digital purchases (e.g., purchases on a wireless Web site), or even bill payments (e.g., Mobile Bill Presentment and Payment).

# M-CONSUMER CONCERNS

A wide range of consumer concerns arise within the m-commerce environment. The main concerns are summarized below:

- **Privacy:** In the information context, privacy refers to a user's fear of other people/organizations knowing what he or she is interested in ("Big Brother syndrome"). Tracking user Internet-browsing behavior and information requests on the wireless Web is a sensitive topic, as it is for its wired counterpart. The ability to know the exact location of a user at all times further escalates the sensitivity of the Big Brother syndrome. Another type of privacy concern for consumers in this area is that their location might be revealed to interested businesses at all times. Knowing the whereabouts of each mobile user may be perceived as threatening to the m-consumer, as this information could be dangerous if intercepted.
- **Security:** Consumer fears regarding the safety of the information exchanged over a wireless network increases with the degree of interaction and the sensitivity of the information exchanged. Security is a critical component in protecting consumer privacy.
- **Reliability:** For any extent of network coverage, it is important that the connection quality be maintained. The inherent concern here is that loss of the connection can result in loss of data (Nielsen, 2000).
- **Download times:** Mobile users should not be forced to spend excessive amounts of time to access desired content (Cole, 2001).
- **Cost:** Users of wired Internet access have the option of subscribing to different transfer rates, which come at different cost levels, subject to their individual needs. Aside from the cost of connecting to the wireless Web, there is also a cost concern for the accessed information.
- **Usability:** Information on the wireless Web should suit not only people's needs but also the medium and the environment. For instance, content needs to be repurposed for mobile devices, so that users can access easy-to-digest pieces of news, not replicated long articles from the wired Web

(McGinity, 2000). This notion ties in with usability, which raises the following questions: How easy is it for the mobile user to access the information sought? What is the quality of the overall experience? Factors influencing the quality of the overall experience include a user's ability to read the screen, input data, manipulate files, and access sites of interest.

- **Content:** Limited content availability is a consideration that prevents customers from accessing the Internet wirelessly. Further user frustration is experienced when they are victims of "walled gardens" (i.e., when they cannot access desired content because it is available only to users of other network carriers).

# MATCHING M-COMMERCE NEEDS AND CONCERNS

The consumer concerns associated with m-commerce identified in the previous section may apply to more than one area of mobile applications. As Table 2 illustrates, the four m-commerce consumer application areas exhibit similar concerns, with cost, security, and privacy prevailing, as they are present as concerns in all application areas. By addressing these three concerns, businesses would reduce consumer reluctance to accept and adopt this new medium.

## The Cost Concern

Who will pay for the content? This is a question that will draw a lot of attention and will require the cooperation of network operators and content providers. For the time being, m-consumers are mostly concerned with connectivity and communication costs. Currently, there are three prevailing pricing options for these services (McGinity, 2001):

- **Flat rate:** A nominal charge for unlimited access for a given length of time (e.g., month).
- **Per minute:** Charged for every minute connected to the network.
- **Per bit:** Charged for the total volume of data transferred in a given period of time.

*Table 2: M-Consumer Needs and Corresponding Concerns*

| Business Application | Concerns |
|---|---|
| Communication | Cost, Privacy, Security |
| Information | Cost, Usability, Privacy, Security |
| Entertainment | Cost, Usability, Download times, Privacy, Security |
| Commerce | Cost, Usability, Security, Privacy |

Adopting a flat-rate pricing model at this stage would be the best approach to lure new customers fast, which is necessary to provide the much-needed critical mass to alleviate the development costs and, in particular, the high license fees for network carries engaged in implementing 3G network technology. The basis for this recommendation lies in the following two observations:

- First, users are accustomed to flat-rate schemes.
- Second, users are in favor of flat-rate schemes because of the model's simplicity and the ability to control expenses.

Once a critical mass is established, different means for pricing may be adopted, and even a combination of models may become available for any particular region, subject to the m-consumer's use of the wireless Web. At that point, pricing based on the data inflow and outflow would be favored by wireless operators, because it would serve as an indirect control on the use of the networks and would help prevent network overload, a situation presently felt by many mobile phone subscribers.

Another dimension to the cost issue is who ends up paying for a wireless interaction in an m-commerce transaction. In North America, both the caller and the receiver of a wireless communication pay their providers for that interaction under current pricing schemes. This scheme represents a significant obstacle to the spread of m-commerce, as consumers will resist having to pay for unsolicited offers received from businesses on their wireless devices. A pricing model, in which the initiator of an m-commerce interaction is responsible for footing the bill, would be a significant boost for consumer involvement in m-commerce activities.

Finally, it is even conceivable that the above models will eventually be replaced by a free, unlimited access, model for the user, subject only to a rental cost for the device, and using m-commerce transaction fees to offset the remaining costs. These fees may be derived from notification services (paid by user), advertising (paid by advertising company), transaction fees on mobile purchasing (paid by merchants, similar to Interac and credit cards), and further means yet to be identified, as the m-commerce market evolves (Simon, 2002).

## The Security Concern

Wireless technology possesses two main vulnerability areas that are a hacker's main attack points. The first point is known as the "Two-Zone problem" or the "WAP Gap." The WAP architecture requires an intermediate gateway (WAP gateway) that encodes and decodes data from the wired encryption format known as SSL (Secure Socket Layer) to its wireless counterpart WTLS (Wireless Transport Layer Security). This process lasts briefly (milliseconds), but the data is unsecured in the interim, as it needs to be decrypted from WTLS into plain text and then reencrypted into SSL. The inherent risk is loss or exposure of data, if a hacker is able to extract the plain text (Gururajan, 2002).

The second point refers to the data stream that is carried through the air medium and is susceptible to "eavesdropping." The success of the hacker in such an attempt depends in part on the encryption algorithm used. The current standard employed by GSM is the A5 algorithm, which utilizes a 54-bit encryption, which is slightly better than the IEEE 802.11 standard RC4-40 algorithm (also known as WEP, or Wired Equivalent Privacy) that only uses a 40-bit encryption. However, both are still not efficient to desired levels (Pesonen, 1999; Bask, 2001). When comparing this level of encryption to the respective levels of wired encryption at 128-bits, it becomes apparent how low the level of wireless security currently is, especially when one considers that hacking a 128-bit encrypted message is also feasible, albeit being rather difficult. In addition, implementing an effective encryption algorithm is further complicated due to the mobile device limitations that still prevail. Limited battery life, low processing memory, and even pricing methods (i.e., per-minute billing), act against the implementation of a 128-bit encryption algorithm in a wireless setting.

Aside from identifying the most likely points of a hack attack, it is important to address the loss or theft of a mobile device as a security issue, because the data stored in the device could be highly sensitive. To combat this situation, mobile users should be empowered through added features for their mobile devices that would safeguard their privacy. These features may be invisible to the user (e.g., memory protection, file access control), or they may require interaction (e.g., log-in software, biometrics) (Gururajan, 2002; Johnson, 2002).

Finally, although security is not synonymous with privacy, it is a critical element in preserving identifiable information as private. As such, privacy concerns arise consequent of the lower security levels of wireless networks and of the potential for using tracking and profiling technologies to offer m-customers unsolicited location-based services, for example. These issues are explored next in some detail.

## The Privacy Concern

Privacy concerns exhibited by m-consumers are similar to those of e-commerce customers. In addition, new privacy concern elements arise consequent of the lower security levels of wireless networks and of the potential for using tracking and profiling technologies to offer m-customers unsolicited location-based services.

The vulnerability of wireless networks creates increased risk for privacy interruptions through potential network security breaches. The ability to snoop in on a user's conversation or even monitor data transmissions generates an uneasiness that the consumer may not be willing to accept. Enhanced security algorithms and hardware improvements can help minimize the risk of such violations.

Positioning services provide additional information companies could use to improve understanding of the mobile user. The ability, however, to know the exact whereabouts of a mobile user may be perceived as threatening by the consumer, as this information could be dangerous if intercepted. Examples of such fears include the following:

- Knowing where mobile users are makes it easier for them to become victims of physical attacks.
- Knowing that the residents of a home are away makes their residence vulnerable.
- Knowing the location of mobile users makes it easier for them to become victims of unsolicited location-based advertising.

The last example, location-based advertising, is one of the most controversial aspects of the ability to track a mobile device and, hence, its user. Companies are using this ability to market their products and services more aggressively. These marketing efforts build on the consumer concern for cost, as they may come at a cost to the mobile user, who may possibly end up paying to read or listen to an incoming advertising message that may be in the form of an e-mail message, SMS, or a phone call.

In effectively addressing the entire range of m-consumer concerns, the active participation of all m-commerce market players is required. The roles and responsibilities for each of these players are examined next.

# ADDRESSING M-CONSUMER CONCERNS

Several companies positioned themselves to play a multifaceted role in this industry, thus creating an entirely new business landscape, where players often have overlapping roles. The mobile value chain becomes more intimate and dynamic, possibly with multiple interactions that do not necessarily preserve a sequential nature and where all market players need to contribute for the industry to reach an optimal level. Thus, a new m-commerce value network (Figure 3) was proposed by Coursaris and Hassanein (2002) that better captures the interactions between the various players in the industry. The mobile value network introduced is made up of customers, network operators, service providers, technology vendors, application developers, and content providers. Because of the multiple interdependencies among value network members, if any of these parties is underdeveloped (or even absent), then the entire network could potentially break down. In addition, each of the six parties identified in this new value network may be made up of additional subsets of companies with more specific business objectives; these possible subsets are identified next, where each value network member is discussed in further detail (Turban, 2002; Kalluvilayil, 2001; Kalakota, 2002; Buckingham, 2000):

- **Customers:** Customers may be the most important value network member, because in the absence of customer demand, there may be little, if any, need for any of the other players in the value network to be present. For example, if the wireless customer does not see the value in nonvoice mobile services made available by content providers (e.g., weather information), then there is little point in network operators maintaining

*Figure 3: M-Commerce Value Network (Coursaris & Hassanein, 2002)*

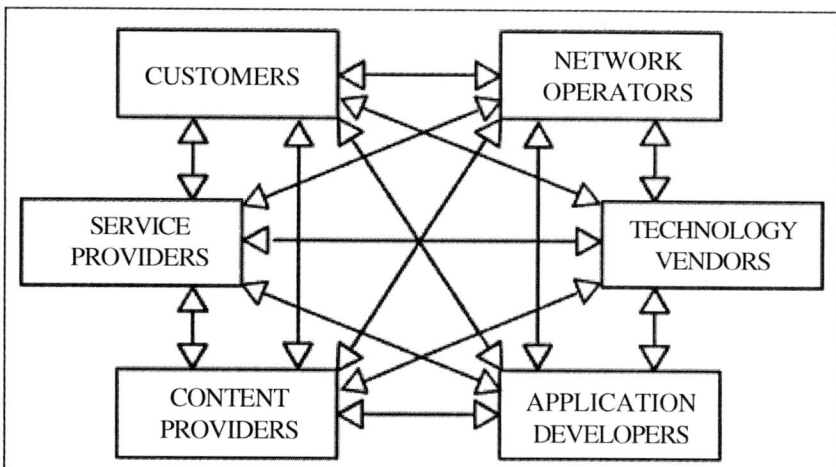

network service (e.g., GPRS), technology vendors manufacturing wireless products (e.g., handsets), service providers offering wireless products and services (e.g., wireless network access), or developers formulating applications (e.g., wireless chat).

- **Network operators:** Arguably the second most significant party after the customer in the m-commerce value network is the network operator (or network carrier). Network operators are crucial in the success of the m-commerce industry, as they are responsible for a wide range of activities. Such activities include deciding if and when to invest in network infrastructure supporting nonvoice services, educating customers about the availability and uses of these new services, and incurring additional expenses to support compatibility with networks of other operators. Such companies typically utilize a subscription fee business model with customers, as well as a transaction-based fee (e.g., per "hit") business model with content providers.

- **Application developers:** Application developers include software developers and systems integrators that provide a wide range of services, such as hosting and transaction processing. Ultimately, these companies are responsible for delivering a practical solution for customers enabled through available technology. Thus, if they are successful in identifying and addressing customer needs, returns will be high for all involved in providing nonvoice mobile services. Developers may offer off-the-shelf products (e.g., chat programs), customized products developed specifically to meet one customer's requirement, or hybrid products based on generic products that are further customized with application-specific data. Typically, the business model adopted by these companies is based on software licensing fees, utility transaction costs, and subscription fees.

- **Service providers (SPs):** Similar to the various Internet Service Providers (ISPs) for the wired Web, Mobile Service Providers (MSPs) emerged to provide an easy way for customers to gain access to wireless networks and available solutions. In addition to this function, some literature includes content providers and operators under this category, as they have come to expand their offerings into the area of servicing customers as well. Strictly speaking, however, MSPs sell products and services of others under their name to customers.

- **Technology vendors:** The mobile value network member that transforms what is desired and theoretically designed to what is actually available is the technology vendor. They supply the necessary hardware and some of the software to enable the convergence of telecommunications and IP networks, ranging from transmission towers to mobile handset receivers. Internally, this group is made up of companies concentrating on different aspects of infrastructure; these further classifications can be seen in Figure 3, where the value network members are titled "technology platform vendors" (e.g., Palm and Microsoft), "infrastructure and equipment vendors" (e.g., Alcatel and Ericsson), "application platform vendors" (e.g., IBM and Motorola), and "handset vendors" (e.g., Palm and Compaq). These groups need to coordinate their efforts to prevent market inefficiencies, such as delays in releasing appropriate handsets for the latest networks made available (e.g., the case with WAP-enabled handsets). Such inefficiencies can cause not only financial turmoil for some of the players but also even complete abandonment and failure of new technology initiatives. The typical business model is based on sales or leasing, as well as license and maintenance fees applicable for software.

- **Content providers:** The information a customer accesses when using the wireless Web may be made available through content providers (e.g., Reuters), content aggregators (e.g., digitallook.com), or portal providers (e.g., Yahoo!). For simplicity, these threes types (or subsets) of companies are grouped here as "content providers." The typical business model is based on advertising and subscription fees. Content providers in the mobile industry currently tend to enter into exclusive agreements with network operators, giving rise to what is known as the "walled garden," where subscribers to specific network carriers gain access to an exclusive set of content providers. This is a symptom being addressed in efforts to provide a truly ubiquitous wireless network that is not only technologically compatible but also offers unrestricted access to content to all mobile users, regardless of carrier selection.

Revisiting the mobile value network while bearing in mind the m-consumer's needs and concerns for business applications would highlight the areas for which each of the value network members is responsible. A summary of these responsibilities is given in Table 3. While this summary is not exhaustive, it highlights the most pressing areas for consumers and the actions necessary to be taken by each of the value network members. Through the aggregated progress of these market players, m-commerce has the potential of realizing its potential growth in the m-consumer segment.

# M-COMMERCE CONSUMER BUSINESS APPLICATIONS

In this section, various business applications targeting the mobile consumer will be reviewed, and how they address the interdependence of the three areas already discussed in this chapter (i.e., wireless technology, m-consumer interaction, m-consumer needs/concerns) will be discussed. (This discussion is summarized in Table 4.) The characteristics identified for each business application, in the table, include the following:

- Consumer needs addressed by the business application (identified through this research).
- Interaction modes covered by the business application (based on Figure 2).
- Global market size (in users), unless otherwise noted, if available (sources are referenced in the table).
- Perceived value and willingness to pay for the business application (Daum, 2001; Wong, 2001).
- Concerns associated with the business application (referring to section "Matching m-commerce needs and concerns").

## Table 3: Mobile Value Network Member Responsibilities to M-Consumer

| m-Commerce Value Network Members | Cost | Privacy | Security | Usability | Reliability | Download Times | Content Availability |
|---|---|---|---|---|---|---|---|
| **Network Operators** | Offer network access at reasonable rates | Disclose & enforce a strong privacy policy | Implement latest network security measures | Implement networks supporting features enhancing usability through protocols and bandwidth | Maintain high network reliability | Enhance / optimize networks to support high transfer rates | Implement networks supporting rich content; Offer incentives to content providers |
| **MSPs** | Offer products & services at reasonable rates | Disclose & enforce a strong privacy policy; Seek TTP approval | Endorse latest network security measures; seek TTP approval | Develop portals with high degree of usability | Maintain high system reliability; Seek TTP approval | Enhance / optimize systems to support high transfer rates | Create portal with large content base; Offer incentives to content providers |
| **Technology Vendors** | Offer products at reasonable rates | Offer technology enhancing privacy in products | Implement latest device security measures | Develop devices with high degree of usability | Develop products with high reliability | Develop products supporting high transfer rates | Develop products supporting rich content |
| **Application Developers** | Offer applications at reasonable rates | Offer measures to help support privacy protection in applications | Implement application security measures | Develop applications with high degree of usability | Develop applications with high reliability | Develop applications supporting high transfer rates | Develop applications supporting rich content |
| **Content Providers** | Provide content at reasonable rates | Disclose & enforce a strong privacy policy; Seek TTP approval | Secure websites | Develop websites with high degree of usability | Develop websites with high reliability | Optimize web content for fast download | Constantly generate new content of interest |

*Table 4: Characteristics of M-Commerce Consumer Business Applications (Coursaris & Hassanein, 2002)*

| Business Application | Needs[1] | | | | Interaction Mode | User market in millions, 2005[9] | Value | Concerns | Technology Requirements[8] |
|---|---|---|---|---|---|---|---|---|---|
| | 1 | 2 | 3 | 4 | | | | | |
| **Communication** | | | | | | | | | |
| - Voice | Π | Π | Π | Π | $W_{B2C}$ $W_{C2C}$ | 1268 | Highest | Cost, Privacy | **1G** / **2.5G**, Voice module |
| - SMS | Π | Π | Π | Π | $W_{B2C}$ $W_{C2C}$ | 1268 | Highest | Cost | **2G** / **2.5G**, WAP 2.0 |
| - e-Mail | Π | Π | Π | Π | $W_{B2C}$ $W_{C2C}$ | 200[A] (by 2004) | Highest | Cost | **2G** / **2.5G**, WAP 2.0 |
| - Data Transfer | Π | Π | Π | Π | $W_{B2C}$ $W_{C2C}$ $W_C^2$ | 2.8 (residential) 9.5 (Total) | Highest | Cost | 2.5G / 3G |
| **Information** | | | | | | | | | |
| - Web browsing | Π | Π | Π | | $W_{B2C}$ | 614 | Highest | Cost, Usability | **2G** / **2.5G** / **3G**, WAP 2.0 |
| - Traffic/Weather | | Π | Π | | $W_{B2C}$ | N/A[6] | Highest | Privacy, Usability | **2G** / **2.5G**, LBS[7] |
| **Entertainment** | | | | | | | | | |
| - Gaming | | | Π | Π | $W_{B2C}$ $W_{C2C}$ | 775 (Total) 200[B] | Highest | Cost, Usability | **2G** / **2.5G** / **3G**, WAP 2.0 |
| - News/Sports | | Π | Π | Π | $W_{B2C}$ | N/A | High | Cost, Usability, Privacy | **2G** / **2.5G** |
| - Downloading Music/Video/Img. | | | Π | Π | $W_{B2C}$ | N/A | Medium | Download times, Cost | 2.5G / 3G, WAP 2.0 |
| - Horoscope/ Lottery | | Π | Π | Π | $W_{B2C}$ | N/A | Low | Cost, Privacy | **2G** |
| **Commerce** | | | | | | | | | |
| - Ticketing (e.g., Event, Cinema) | Π | | | | $W_{B2C}$ | N/A | Highest | Cost, Usability, Security, Privacy | **2G** / **2.5G** |
| - Pre-Payment | | | | Π | $W_{B2C}$ | 18.3 (by 2003) | Highest | Security | **2G** / **2.5G**, Real-time Billing |
| - Banking | | Π | Π | Π | $W_{B2C}$ | 798 | High | Security, Privacy | **2G** / **2.5G** |
| - Advertising | | Π | | Π | $W_{B2C}$ | $16-23 billion | Medium | Privacy (Spam) | 2.5G/3G, LBS, WAP 2.0 |
| - Retailing | | Π | | Π | $W_{B2C}$ | 469 | Medium | Security, Privacy, Usability | 2.5G / 3G, LBS, WAP 2.0 |

- Technology requirements for the business application (supported in the remaining discussion).

The applications presented in the table are those of highest interest to consumers, according to research (Daum, 2001; Wong, 2001), and they often address multiple needs. For example, mobile banking would include options to access a user's account to obtain a balance, transfer funds, and even proceed with trading securities. This application, therefore, satisfies both the need to access information as well as engage in commercial transactions. In general, applications have been grouped under a need area in the first column of Table 4, according to which need they predominantly cater.

## Communication Applications

Satisfying communication needs represents the foundation for fulfilling all the remaining m-consumer needs. From the "Interaction Mode" column in Table 4, it is evident that only communication applications target all of the different consumer interaction modes. As such, these applications can cater to a wider audience, whose members appear to be more interested in and more willing to pay for this type of application.

Cost appears to be the primary concern and would thus require network carriers to revisit their pricing models, and consequently come up with various options (i.e., subscription, pay-per-use) in an attempt to satisfy different consumer preferences. Finally, with respect to technology, only "data transfer" is affected by the slow adoption of 2.5G, because voice, SMS, and e-mail can operate efficiently within existing technologies. Future enhancements exist in VoiceXML, the technology that will enable voice-driven applications, some of which are already available (e.g., speaking out the name of the person whose phone number is to be dialed). Enhancements are also expected in 3G networks and the WAP 2.0 protocol, which will support rich content in SMS and e-mail communications, as well as provide for higher transfer rates for data transfers.

## Information Applications

These applications target the wireless B2C consumer interaction mode. A high consumer interest in these wireless information Web sites suggests an opportunity for content providers to start charging mobile consumers for their services (i.e., subscription, pay-per-use), if they are not doing so already.

Cost and usability take front stage in terms of m-consumer concerns, and along with network carriers and content providers rethinking their pricing models, content providers need to ensure a high level of usability to avoid customer dissatisfaction and potential market attrition.

Finally, 2.5G and 3G network technologies will help improve the wireless Web experience, and the available information could become rich in form, yielding higher customer appreciation and interest. In addition, future location-based services could enable dynamic searching and comparison for location-specific information.

## Entertainment Applications

Entertainment can involve various activities, some of which can satisfy various types of m-consumer needs. Gaming currently appears to be the hottest segment, with emphasis on the teenage and young adult communities (Ovum, 2001). Until recently, however, this offering was limited due to protocol constraints (WAP did not allow for graphics and rich content). The next generation of protocols should be able to address this problem. As usual, cost and usability are in the foreground as concerns, along with download times. For cost, downloads can be purchased individually or through a subscription, giving m-consumers added flexibility. A mobile user may seek entertainment for a short interval on a spontaneous basis. Therefore, excessive download times will not be well received. Finally, 2.5G and 3G network technologies, along with the introduction of WAP 2.0, will help improve not only the gaming experience but also other entertainment-related applications (Harmer, 2001).

## Commerce Applications

Although gaming appears to be the short-term cash cow for m-commerce, mobile banking presents the primary application for generating the much-needed critical mass in the near future, which in turn, can yield significant revenues. In addition, banking is an application that is not a passing fad and is not subject to the latest video and audio technologies; rather, it is an important provision for mobile consumers and their need to save time from routine activities, such as going to the bank to pay a bill. Mobile banking is a key application for supporting the mobile payment mechanisms needed for other m-commerce applications to take place.

Cost and usability are present concerns once again, but due to the sensitive nature of the information exchanged in a commercial transaction, security and privacy concerns prevail. The limitation in addressing these concerns effectively today lies in the existing infrastructure. The two main points of potential hack attacks were identified earlier in the chapter and were found to be the "Two-Zone problem" or the "WAP Gap" as well as "eavesdropping." The WAP Gap can be addressed effectively in devices accessing GSM networks, as these devices handle the conversion from WTLS to SSL internally on the SIM (Subscriber Identity Module) card and, therefore, minimize the risk of a hack attack and improve overall performance as airtime required for conversion is reduced. Other options are explored through new technologies, including WIM (Wireless Identification Module) cards that are similar in functionality to SIM cards for non-GSM phones, and J2ME-enabled handsets, which allow the handset to send and receive content directly to and from the HTML server, respectively, without the need for an intermediate gateway (Schwartz, 2000). On the other hand, protection against eavesdropping would require more efficient encryption algorithms (e.g., 128-bit) as well as supporting wireless technology enabling these algorithms (e.g., more powerful wireless devices that are not constrained by battery life or memory). Also, frequently, security features are sidestepped in return for time benefits (mobile users omit or deactivate security features to save on transmission time). Therefore, until security enhancements to wireless technology are in place, mobile users may be reluctant to take advantage of these applications. Finally, m-commerce industry players need to implement sufficient content to serve as incentive for not only converting consumers to mobile users but also retaining these mobile users for the long run.

# GLOBAL M-COMMERCE
# MARKET OVERVIEW

The growing importance of m-commerce is fueled by the phenomenal growth in the wireless market in general. Shown in Figure 4 is the growth experienced in the wireless device market as well as in the subscriber base of wireless Internet services (Morrison, 2001).

According to these forecasts, the global customer base for wireless Internet access is expected to match the overall wireless subscriber base by

*Figure 4: Global Subscriber Base for Wires and Wireless Internet Access (Morrison, 2001)*

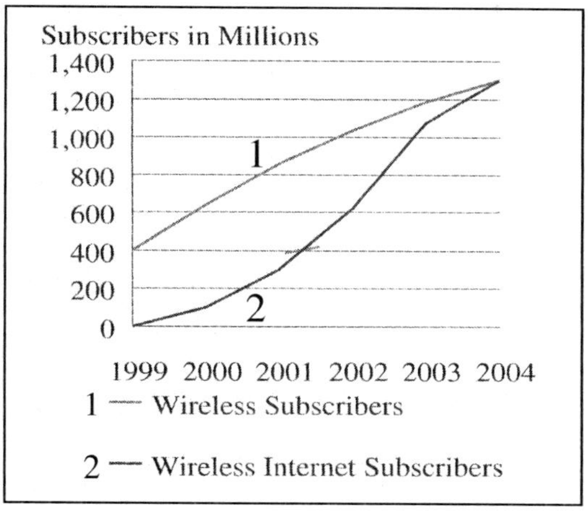

Subscribers in Millions

[Figure: Line graph showing two curves from 1999 to 2004. Y-axis labeled "Subscribers in Millions" from 0 to 1,400 in increments of 200. Line 1 (Wireless Subscribers) rises from about 400 in 1999 to about 1,300 in 2004. Line 2 (Wireless Internet Subscribers) rises from near 0 in 1999 to about 1,300 in 2004.]

1999 2000 2001 2002 2003 2004

1 — Wireless Subscribers

2 — Wireless Internet Subscribers

*Source: Ovum*

2004 (more than 1.2 billion subscribers, or 20% of the world's population) (Morrison, 2001). This represents the number of users who have access to the wireless Internet but may not necessarily be using it.

This growth in wireless Internet subscribers is expected to be matched by a growth in m-commerce-related activities that vary by region, as indicated in Table 5, which outlines forecasted regional m-commerce revenues. These revenue estimates by Jupiter are on the conservative side at $22.2 billion, compared to other research groups that predict m-commerce revenues to be larger by as much as five times (Canvas, 2001). As such, m-commerce represents a market with substantial financial returns, along with additional benefits, such as improved branding and customer service through exploitation of the fast-growing wireless channel. According to Table 5, the fastest growing and largest markets for m-commerce are found in Asia, and, in particular, in Japan, followed by Europe.

A frequently asked question within the m-commerce industry is in regard to the killer application: Is there a killer application, and if so, what is it? Although one will be confronted with many answers, each supported with its respective rationale, a more likely scenario is that killer applications for m-commerce will vary by culture and by country. This is evidenced at this early stage of m-commerce by the varying demand for wireless applications around the world. In Europe, the killer application has been Short Message Service

*Table 5: Regional M-Commerce Revenue (US$ billion) (Canvas, 2001)*

| Region | 2000 | 2001 | 2002 | 2003 | 2004 | 2005 |
|---|---|---|---|---|---|---|
| N. America | 0.01 | 0.1 | 0.2 | 0.7 | 1.8 | 3.5 |
| W. Europe | 0.015 | 0.1 | 0.5 | 1.7 | 4.6 | 7.8 |
| Asia | 0.4 | 1.3 | 2.6 | 5.0 | 7.4 | 9.4 |
| S. America | 0.0 | 0.0 | 0.0 | 0.1 | 0.2 | 0.5 |
| Other | 0.0 | 0.0 | 0.1 | 0.2 | 0.4 | 1.0 |
| **Global** | **0.425** | **1.5** | **3.4** | **7.6** | **14.5** | **22.2** |
| US | 0.01 | 0.1 | 0.2 | 0.6 | 1.7 | 3.3 |
| Japan | 0.4 | 1.2 | 2.1 | 3.5 | 4.5 | 5.5 |

(SMS); in Japan, interactive games and pictures; and in North America, e-mail via two-way interactive pagers (e.g., RIM's BlackBerry) plus WAP-based wireless data portals providing news, stocks, and weather information.

These "killer applications" will take on many forms as the wireless networks and devices evolve, always improving network connectivity, device form factors, and capabilities. One expectation is that content, services, and applications for wireless devices will become increasingly available.

An overview of the three leading markets for m-commerce is provided next, in an attempt to understand the variations that exist in the adoption of m-commerce services in these different geographic markets.

## Europe

With GSM being the single digital mobile telecommunication standard implemented, European countries are positioned well in their race to mobility. GSM has international roaming capability, it is supported in over 159 countries, and it accounts for over 64% of the world's wireless market (Evans, 2001). Consequent of the uniformity in network implementations in this region, market penetration is highly feasible, as interoperability issues are minimized. The last hurdle that network carriers will have to overcome is the estimated cost of US$80 billion necessary to upgrade the existing digital phone networks to accommodate 3G (Dorey, 2002).

To generate much revenue for the continuous upgrade efforts, as well as to alleviate the cost from the high license fees, the availability of wireless applications that are in demand becomes critical. According to the GSM Association, in Europe, SMS has been one such wireless application, for which demand exceeded 50 billion global text messages sent within the first quarter of 2001. During the same period in the United Kingdom alone, customers generated 3.5 billion text messages. SMS has been popular not only for messaging between wireless users but also as a marketing medium through which m-consumers can respond to television shows (e.g., MTV) that encourage audience participation. Due to the above demand for SMS, it has been coined as the "killer app" for Europe (Evans, 2001).

Extending from the communication applications, Europe also launched initiatives with transactional capabilities for commerce. One example is the m-shopping service provided by the Safeway grocery chain. Through PDAs provided by Safeway, m-consumers can create shopping lists and submit orders. These orders are then sent to the store, where the staff is responsible for collecting and packaging the purchased items for the customer to pick up (Evans, 2001).

In general, Europeans are receptive to m-commerce due to various factors. An industry-related factor involves the pricing strategy adopted by network carriers, who implemented a "caller pays" model for voice communication. Cultural factors include a "café" culture (i.e., a tendency to be active away from home), as well as patience in using many keystrokes in generating text messages and e-mails on their wireless phones (Dorey, 2002). Finally, with an effective transportation system and relatively high gas prices, Europeans end up spending a lot of time on public transportation, with time on their hands. This provides them with strong motivation for using the wireless devices to invoke m-commerce services. Together, the above factors result in a higher acceptance of m-commerce.

## Asia-Pacific

One of the biggest success stories for the wireless industry has come from the Asia-Pacific region. NTT DoCoMo's i-mode is Japan's largest network carrier, with more than 30 million customers, capturing 59% of the domestic wireless market (NTT, 2002). Since i-mode's launch in 1999, the respective adoption rate and revenues generated have been the envy of wireless carriers around the world. This growth has been due, in part, to NTT DoCoMo's exclusive offering through i-mode, as competing network carriers operating in WAP could not offer comparable wireless content and services. Using packet data transmissions, fees for wireless services are charged by the amount of data transmitted and received rather than the amount of airtime.

Pioneering many of the wireless services worldwide, i-mode users have access to mobile banking, travel reservations, restaurant and town information, messaging services for news, i-mode-compatible Web sites, e-mail, entertainment sites, and downloadable ring tones. Some of the content is made available by DoCoMo at no cost to the user, while other content is subject to a monthly fee that ranges from 100 to 300 yens per month per offering.

The popularity of wireless services helped maintain a momentum that brought 3G to the Japanese market in the third quarter of 2002. With nearly 150,000 mobile phones being ordered, only 4500 were actually given out in this initial phase of "FOMA," which stands for "Freedom Of Mobile multimedia Access." Of these phones, 1200 were equipped with a video screen to facilitate some of the 3G-supported wireless applications, such as video playback (Evans, 2001).

In general, the Japanese culture has embraced technology, and along with the low-cost alternative of accessing the Internet (aided by a lower PC penetration compared to Western Europe and the United States) and charging based on data volume for wireless services, m-commerce has been very successful in Japan and is experiencing growth in other countries within this region (in particular, China and Hong Kong) (Dorey, 2002). Finally, with a similar transportation situation to that of Europeans, Japanese consumers end up spending a lot of time on public transportation with time on their hands. This provides them with a strong motivation to use the wireless devices to invoke m-commerce services.

## North America

The United States has been leading the way in adoption of wireless technologies in this region, followed by Canada, who is catching up on various levels (e.g., m-commerce growth, penetration of wireless devices). Overall, however, North America has been known to lag behind Japan and Europe in its m-commerce efforts. Several factors contribute to the slow adoption of m-commerce in North America. One of these factors has been interoperability, as currently there are six U.S. and four Canadian national network carriers that operate on different network standards (Dorey, 2002). This lack of common network standards will pose an even greater challenge to overcome (in particular, financially) in their efforts to evolve to a 3G environment. Furthermore, high PC penetration rates and a low-cost alternative for voice communications offered through wired telecommunications service do not create a desirable m-commerce-value proposition for consumers.

Culture appears to be acting as a barrier for m-commerce as well. North Americans, in general, exhibit a "stay-at-home" social outlook. Also, there has been a demonstrated lack of patience and an expressed dissatisfaction with the usability of currently available wireless devices (e.g., challenging to generate e-mails on a wireless phone). Finally, little knowledge of wireless technology and applications by the average North American consumer poses yet another challenge for m-commerce adoption (Dorey, 2002). Another factor contributing to the slow adoption of m-commerce in North America could be the affinity of North Americans to driving their own cars. The public transportation infrastructure is inferior to both Europe and Japan. The cost of owning and operating a vehicle is lower compared to that in Europe and Japan. All of these factors combine to instill a driving culture in North America. For example, because most people drive to and from work every day, they are not in a position to use mobile devices for m-commerce services during their commute, as easily as those using public transportation.

Consequent of the above issues, the wireless consumer market has been small. However, the business market has been more responsive. In particular, one of the major trends in the United States has been the use of Research In Motion's (RIM) wireless devices for receiving and sending corporate e-mail. Through an always-on service for wireless e-mail using the DataTAC and Mobitex wireless networks, many companies adopted this technology to further enable their organizations. In addition, wireless Web access via WAP-enabled cell phones is a growing application in the United States (Dorey, 2002).

# DISCUSSION AND FUTURE TRENDS

The m-commerce industry is fast growing, with estimates of reaching a user base of 1.3 billion people around the world by 2004 (Morrison, 2001), contributing to an overall market in excess of US$22 billion (Canvas, 2001). Industry players, ranging from network carriers to content providers, hope to capture part of this revenue. However, early results were not up to the hyped expectations due to a combination of reasons ranging from technology limitations or limited business applications. Concerns center on the issues of cost, privacy, reliability, download speed, usability, security, and content availability. For m-commerce to reach its potential, these concerns will have to be effectively addressed, and collaboration among all value network members will be essential. It should also be noted that health concerns, although not linked to any particular application, pose another barrier for adoption of wireless technology. On this issue, the m-commerce industry will need to clearly communicate any findings, so as to reduce fears of health hazards consequent of mobile device usage.

For the most part, the drawbacks found in using mobile devices for Web-based functions will be resolved in the near future, as advancements are being made simultaneously in wireless networks, wireless protocols, mobile devices, and supporting technologies.

The traditional categories for wireless devices include wireless phones, wireless PDAs, and wireless laptops. The latest innovation in wireless devices involved the introduction of smart phones, which represent wireless devices that enabled users with a new set of capabilities, derived from more than one of the traditional wireless devices. This convergence trend is expected to continue in the foreseeable future to support consumer demands for mobile devices that can provide a wider range of capabilities (Keyte, 2001).

With respect to wireless communication protocols (e.g., WAP, i-mode), it is unlikely that any one will prevail over the others on a global basis. The more likely scenario will be that wireless devices will evolve to support all protocols seamlessly. This is one of the goals set to be achieved with the implementation of the 3G wireless networks (hence, the name UMTS). Wireless networks will continue to be implemented that offer higher bandwidths and, consequently, can support rich content, such as streaming video, at faster download times than those available today. Japan is leading the rest of the world, having implemented their 3G network in the fourth quarter of 2001 and having already announced that 4G is expected to arrive in 2006 (NTT, 2002). Although 4G will provide higher bandwidth than 3G, it is the latter that will help address the main problem of ubiquity. Ubiquity is a critical success factor for m-commerce, and with the whole world eventually migrating to 3G, there will be no more barriers to prevent anytime, anywhere m-commerce.

One area that deserves particular attention is related to content management. Issues in this area arise from the lack of compatibility and the absence of automated translation mechanisms between the wired and wireless Web environments. It may be the case that before long, language interpreters or translators will convert a single Web site to any standard, taking into consideration the form factor involved. For now, these applications are still emerging, and organizations are required to go through the nuisance of running two separate sites (i.e., one for the wired Web and one for the wireless Web) and managing the associated complexities. Consequently, additional resources are required that are estimated at 30% above the cost of implementing an HTML Web site (Little, 2001).

Once technology-related problems are addressed effectively, the emphasis for market players will shift to developing content and implementing effective m-commerce business models. Understanding the needs and wants of the m-consumer, as outlined in this chapter, can facilitate creation of a loyal m-consumer base. Businesses targeting m-consumers need to understand that a Web-enabled mobile device does not necessarily guarantee that a user will take advantage of this capability. Currently, the success story for m-commerce comes from the Far East, where Japan successfully captured 30 million users in less than 3 years on its i-mode platform. This success is largely due to the content that was made available early on, an element that was not present for WAP users in other regions (Levy, 2001). Development language and protocol limitations were partly responsible for this situation, but with WAP 2.0 addressing most of these concerns, content providers need to take charge and give users something to go mobile for, other than communicating. "Content is king" may be an old cliché, but it holds true for this phase of m-commerce, where users do not see a limitation of devices but rather one of content, and are, therefore, reluctant to make the transition to the wireless Web.

Asian countries, in particular, Japan, are expected to continue to dominate the m-commerce market in the near future, although the rest of the world, in particular, Western Europe, is closing the gap. It will be interesting to see how underdeveloped countries will respond to the m-commerce opportunity, which can act as a "leapfrog" technology. M-commerce can reduce the digital divide between developed and underdeveloped countries by allowing underdeveloped countries to implement wireless networks that will serve as their main communications infrastructure.

Future research in the area will be focused on issues related to devising m-commerce business models that can take full advantage of the fast unfolding technological improvements in the areas of wireless networks, devices, and protocols. Developers of such models will have to pay close attention to satisfying the needs of m-consumers while minimizing their concerns. Another area of key importance for future research in this field is the usability of mobile devices and m-commerce Web sites, because it highly impacts the rate of adoption of m-commerce activities by m-consumers.

# ENDNOTES

1. TDMA: Time Division Multiple Access; CDMA: Code Division Multiple Access; GSM: Global System for Mobile Communications.
2. HSCSD: High-Speed Circuit-Switched Data; GPRS: General Packet Radio Service; EDGE: Enhanced Data for GSM Evolution.
3. W-CDMA: Wideband CDMA; UMTS: Universal Mobile Telecommunication System.
4. D-AMPS: Digital AMPS; CDPD: Cellular Digital Packet Data; PDC: Personal Digital Cellular.
5. 1. Communication; 2. Information; 3. Entertainment; 4. Commerce.
6. N/A = Not Available.
7. LBS = Location-Based Service.
8. Available technology is in bold, while future technology is shown in normal font. (Note: 2.5G is available but is not yet widely used.)
9. Source: allNetDevices, http://www.canvasdreams.com/viewarticle. cfmarticleid=941; except A = ARC Group, 1999, http://www.epsltd.com/ IndustryInfo/Statistics/mobilestats.htm, and B = Datamonitor, 2000, http: //cyberatlas.internet.com/markets/wireless/article/0,,1009 4_455141,00.html.

# REFERENCES

3GAmericas. (n.d.). *TDMA Standards Overview*. Retrieved October 15, 2002, from the World Wide Web: http://www.3gamericas.org.

Bask, J. (2001). Pervasive computing: Travel and business services. Telecommunications Software and Multimedia Laboratory. Retrieved October 15, 2002, from the World Wide Web: http://www.tml.hut.fi/Studies/Tik-111.590/2001s/papers/joni_bask.pdf.

Buckingham, S. (2000). Yes 2 GPRS. *Mobile Streams*. Retrieved May 31, 2002, from the World Wide Web: http://www.mobilewhitepapers.com/pdf/gprs.pdf.

Canvas Dreams. (2001). *Global market statistics for mobile commerce, Part 2*. Retrieved June 15, 2002, from the World Wide Web: http://www.canvasdreams.com/viewarticle.cfm?articleid=943.

Cole, C. (2001). 5 things I want from my mobile. *m-Commerce World*. Retrieved June 15, 2002, from the World Wide Web: http://www.internetworld.co.uk/mcomm/vRoot/articles/article.cfm/B6D4ACE6-D1D4-11D4-BEE900B0D0A143DF.

Coursaris, C. & Hassanein, K. (2002). Understanding m-commerce: A consumer centric model. *Quarterly Journal of Electronic Commerce*, 3(3), 247–271.

Coursaris, C., Hassanein, K., & Head, M. (2003). m-Commerce in Canada: An interaction framework for wireless privacy. *Canadian Journal of Administrative Sciences, 20*, forthcoming.

Daum, A. (2001). Mobile consumers: What do they want? How much will they pay? GartnerG2, (April 2001).

Dorey, S. (2002). Handheld devices and the wireless Internet. The-surfs-up. Retrieved October 15, 2002, from the World Wide Web: http://www.the-surfs-up.com/news/news8p1.html.

Evans, N. (2001). *Business Agility: Strategies for Gaining Competitive Advantage through Mobile Business Solutions*. New York: Prentice Hall.

Gururajan, R. (2002). Mobile computing: Security risks. 23rd World Congress on the Management of Electronic Commerce, Hamilton, Ontario, Canada.

Harmer, J. (2001). 3G products—What will the technology enable? *BT Technology Journal*, 19(1), 24-31.

ITU (International Telecommunication Union). (n.d.). The road to IMT-2000. Retrieved June 3, 2001, from the World Wide Web: http://www.itu.int/imt/what_is/roadto/index.html.

Johnson, D. (2002). Securing your PDA. *IDG.net*. Retrieved June 15, 2002, from the World Wide Web: http://www.idg.net/ic_794581_5056_1-2887.html.

Kalakota, R., & Robinson, M. (2002). *M Business: The race to mobility* (pp. 16–17). New York: McGraw-Hill.

Kalluvilayil, S. (2001). Owning the wireless customer experience. *Edgecom*, July 13, 2001.

Keyte, C. (2001). It's not about the phones! *m-Commerce World*. Retrieved June 15, 2002, from the World Wide Web: http://www.internetworld.co.uk/mcomm/vRoot/articles/article.cfm/A0154418-21C5-11D5-A04E00C04FA0E16A.

Leiner, B., et al. (2002). A brief history of the Internet. *Internet Society*. Retrieved June 15, 2002, from the World Wide Web: http://www.isoc.org/internet/history/brief.shtml.

Levy, S. (2001). Calling the Net. *Newsweek*. Retrieved July 13, 2001, from the World Wide Web: http://www.msnbc.com/news/599997.asp?cp1=1.

Little, J. (2001). M-Commerce. *imazing! CJRW*. Retrieved June 15, 2002, from the World Wide Web: http://www.cjrw.com/imazing/mcommerce.html.

McGinity, M. (2000). Bumpy road ahead for m-commerce. *Inter@ctive Week*. Retrieved June 15, 2002, from the World Wide Web: http://www.zdnet.com/intweek/stories/news/0,4164,2445298,00.html.

McGinity, M. (2001). The Net/wireless meeting of the minds. *Inter@ctive Week*. Retrieved June 15, 2002, from the World Wide Web: http://www.zdnet.com/zdnn/stories/news/0,4586,2457813,00.html.

Morrison, D. (2001). Technology push and customer pull: The wireless Internet comes of age. Presentation at McMaster University, Hamilton, Ontario, Canada.

Nielsen, J. (2000). *Designing Web usability: The practice of simplicity*. Indianapolis, IN: New Riders Publishing.

NTT DoCoMo. (2002). Retrieved October 15, 2002, from the World Wide Web: http://www.nttdocomo.com.

Peck, A. (2001). WAP's summer of discontent. *m-Commerce World*. Retrieved June 15, 2002, from the World Wide Web: http://www.internetworld.co.uk/mcomm/vRoot/articles/article.cfm/87DB2C1B-D4FC-11D4-A9E300C04FA0E16A.

Pesonen, L. (1999). GSM interception. Telecommunications Software and Multimedia Laboratory. Retrieved October 15, 2002, from the World Wide Web: http://www.tml.hut.fi/Opinnot/Tik-110.501/1999/papers/gsminterception/netsec.html.

Pocket Directory. (2001). Smart phones. Retrieved October 15, 2002, from the World Wide Web: http://www.pocketdirectory.com/hardware/hproducts.aspx?idCat=4&selHId=1.

Schwartz, E. (2000). Fixing a security hole when the rain gets in: Two-zone encryption limits wireless usage. *InfoWorld*. Retrieved October 15, 2002, from the World Wide Web: http://www.infoworld.com/articles/op/xml/00/12/04/001204opwireless.xml.

Simon, H. (2000). Sinking your teeth into m-commerce. *Intelligent Enterprise*. Retrieved October 15, 2002, from the World Wide Web: http://www.intelligententerprise.com/000818/supplychain.shtml.

Turban, E., Lee, J., Warketin, M., & Chung, M. (2002). *Electronic Commerce: A Managerial Perspective, 2002* (p. 867). New York: Prentice Hall.

WAP Forum. (2001). WAP 2.0 Technical White Paper. Retrieved June 15, 2002, from the World Wide Web: http://www.wapforum.com/what/WAPWhite_Paper1.pdf.

Wong, R. & Jesty, R. (2001). The wireless Internet: Applications, technology, and market strategy. ARC Group, September 10, 2000. Retrieved June 6, 2002, from the World Wide Web: http://www.arcgroup.com/press/rel_wirelessint.htm and http://www.allnetdevices.com/wireless/opinions/2001/01/10/the_wireless.html.

**Chapter VII**

# Intelligent Product Brokering and User Preference Tracking

Sheng-Uei Guan, National University of Singapore, Singapore

Chon Seng Ngoo, National University of Singapore, Singapore

Fangming Zhu, National University of Singapore, Singapore

## ABSTRACT

*One potential application for agent-based systems has been in the area of m-commerce. In most current systems, user-supplied keywords are normally used to generate a profile for the user. In this chapter, a design for an evolutionary ontology-based product-brokering agent for m-commerce applications is proposed. It uses an evaluation function to represent the user's preference instead of the usual keyword-based profile. By using genetic algorithms, the agent tries to track the user's preferences for a particular product by tuning some parameters inside. A prototype was implemented in Java, and the results obtained from our experiments look promising.*

# INTRODUCTION

One potential application for agent technology is in the area of mobile commerce (m-commerce) (Nwana & Ndumu, 1996; Aylett et al., 1998). According to a study done by Frost and Sullivan,[1] it was projected that electronic commerce (e-commerce) conducted via mobile devices such as cellular phones and Personal Digital Assistants (PDAs) will become a whopping $25 billion market worldwide by 2006. Some of the driving factors behind m-commerce were attributed to the compactness and high penetration rate of these mobile devices.

However, despite all the hype and promises about m-commerce, several main issues will have to be resolved before agent technology can be fully adopted into any m-commerce system (Nwana & Ndumu, 1997; Morris & Dickinson, 2001). Clumsy user interfaces, cumbersome application, low speeds, flaky connections, and expensive services soured many who tried m-commerce. Security and privacy concerns also dampened enthusiasm for m-commerce.

Taking these concerns into account, it seems like good old e-commerce will remain as the preferred choice for online transactions for many years to come. Customers will only use wireless mobile devices to access the Internet if they have a good reason to do so. Therefore, in order to entice customers to participate in m-commerce, the developers will have to offer something that is unique and which no self-respecting consumer can live without. One of the potential "killer" applications for m-commerce could be an intelligent program that is able to search and retrieve a personalized set of products from the Internet for its user.

Currently, when a user wants to search for a particular product on the Internet, what he will normally do is use popular search engines, such as Altavista,[2] and enter keywords that describe the product. These search engines will process these keywords and generate a large number of links for the user to visit.

Although these are the more common methods of searching for information on the Internet, they may not be the best or the most efficient. Neither the search engine nor the Web site know the preference of the user and, hence, might provide information that is irrelevant to the user. For example, if the user wants to search for information about "mobile agents," the search engine could return links to "insurance agents" instead.

In agent-based m-commerce, agents act on behalf of their users by carrying out delegated tasks automatically. Currently, there is no single agent that can perform all the tasks meted out by the user effectively. Like humans, specialized agents are required that are able to work in a specific type of environment. A product-brokering agent seems to be a potential solution for this scenario. The agent will search for the products in the background, with minimal user intervention, thereby allowing the user to concentrate on other tasks. It could be programmed with the user's preferences in mind and filter out irrelevant products automatically. The agent could also detect shifts in the user's interest and adjust accordingly to suit the user.

Described in this chapter is the design of an intelligent ontology-based product-brokering agent capable of providing a personalized service for its user. It does this through *user profiling* (Soltysiak & Crabtree, 1998a). Such agents are able to learn user preferences over time and recommend products that might interest the user. This technique has been used successfully for certain types of agent tasks, typically those that are information intensive and often involve the World Wide Web.

# RELATED WORK

Personalized product-brokering agents require a profile of the user in order to function effectively. The agent would also have to be responsive to changes in the user's interests and be able to search and extract relevant information from outside sources.

At MIT Media Labs, Maes and Sheth (Maes, 1994; Sheth & Maes, 1993) have come up with a system to filter and retrieve a personalized set of USENET articles for a particular user. This is done by creating and evolving a population of information-filtering agents using genetic algorithms.

Some keywords will be provided by the user, and they represent the user's interests. Weights are also assigned to each keyword, and the agents will use them to search and retrieve articles from the relevant newsgroups. After reading the articles, the user can give positive or negative feedback to the agents via a simple graphical user interface (GUI). Positive feedback increases the fitness of the appropriate agent and also the weights of the relevant keywords (vice versa for negative feedback). In the background, the system periodically creates new generations of agents from the fitter species, while it eliminates the weaker ones. Initial results obtained from their experiments showed that the agents are capable of tracking its user's interests and recommend mostly relevant articles.

While the researchers at MIT require the user to input their preferences into the system before a profile can be created, Crabtree and Soltysiak (Crabtree & Soltysiak, 1998; Soltysiak & Crabtree, 1998b) believed that the user's profile can be generated automatically by monitoring the user's Web and e-mail habits, thereby reducing the need for user-supplied keywords.

Their approach is to extract high information-bearing words, which occurs frequently in the documents that are opened by the user. This is achieved by using ProSum,[3] which is a text summarizer that can generate a set of keywords to describe the document and can also determine the information value of each keyword. A clustering algorithm is then employed to help identify user's interests, and some heuristics are used to ensure that the program could perform as much of the classification of interest clusters as possible.

However, they have not been completely successful in their own experiments. The researchers admitted that it would be difficult for the system to classify all the user's interests without the user's help. Nevertheless, they believed that their program has taken a step in the right direction by learning user's interests with minimal human intervention.

A new product-brokering agent usually does not have sufficient information to recommend any products to the user. Hence, it has to get product information from somewhere else. A good source of information will be the Internet. In order to do that, a method suggested by Pant and Menczer (2002) is to implement a population of Web crawlers called *InfoSpiders* that search the WWW on behalf of the user. Information on the Internet will be gathered based on the user's query and then will be indexed accordingly.

These agents initially rely on traditional search engines to obtain a starting set of URLs, which are relevant to the user's query. The agents will then visit these Web sites and decode their contents before deciding where to go next. The decoding process includes parsing the Web page and looking at a small set of words around each hyperlink. A score is given based on their relevance to the user. The link with the highest score is then selected, and the agent visits the Web site.

# DESIGN OF PRODUCT-BROKERING AGENT

A product-brokering agent can be used to search for all kinds of products. In our application, the agent will be used to search for some computer products, namely, *CPU*, *Mainboard*, and *Memory*. It is possible to extend the application to search for other products.

Similar to the information-filtering agents by Sheth and Maes, an initial population of product-brokering agents will be created and evolved using some form of genetic algorithm. However, in this design, the profile of the user is not based on any keywords supplied by the user. In fact, no keywords are required to be entered by the user. Instead, each agent will have an evaluation function that will be used to calculate the value of each product. Products that have a higher value will have a higher chance of being recommended by the agent. This evaluation function has some tunable parameters that characterize the user's preferences for a particular category of products.

In this design, some assumptions were made about the system. One important assumption is that the user is a rational person and will select a product rationally. Another assumption is that the value that a user places on a product can be calculated. The product values that we are focusing on will be those that can be calculated by using some tangible attributes (e.g., price) of the product. The agent will not be able to calculate the intangible value (e.g., branding) that a user placed on the product. If these assumptions are not met, the agent will not be able to track the user's preferences successfully.

## Ontology

Before the product-brokering agent is able to explore the Internet and retrieve product information for the user, the agent needs to have some prior knowledge, such as the URL of some relevant Web sites, keywords, or some quantifiable attributes that can be used to describe the product.

It could be tedious if the user has to enter such information into the agents when he wants to search for a particular product. An alternative to this is to create a *product ontology*, as shown in Figure 1, that already contains some information (e.g., URL of relevant Web sites). Creating a product ontology involves defining the meaning of each term that is used to describe the product, the valid range of values, and the relationship between one another.

The product ontology was implemented in a tree-like structure, with the leaf nodes representing the products and the parent node representing the product category. Each leaf node actually contains a class called *productInfo*, which has some prior information about the product. New products can easily be added as a leaf node to the parent node. When the leaf node is selected, it will pass the product information to the product-brokering agents.

*Figure 1: Screenshot of Product Ontology*

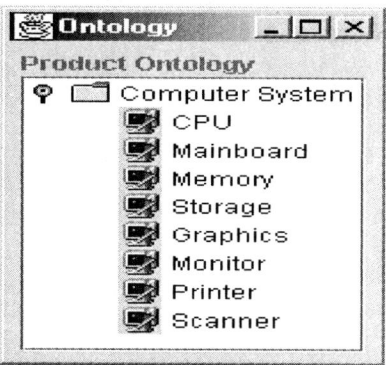

## Product-Brokering Agent

After describing how the agents are going to obtain their product knowledge, the next stage is to define the agent. A unique agent name will be given to each agent so that we can identify and differentiate the agents from one another.

## Agent's Fitness Function

To calculate the fitness of the agent, the proposed fitness function was defined by using the following equation:

$$Fitness = \frac{\sum_{n=window\_size} points \ earned \ in \ the \ recent \ n \ generations}{n} \qquad (1)$$

This fitness function is basically an unweighted (or simple) moving average of the agent's fitness, where $n$ is the window size of the moving average. Using Equation (1), the agent's fitness is obtained by averaging the number of points earned by the agent in the current and the previous $n$-$1$ generations. By varying the value of $n$, we can effectively control the number of generations under consideration, and the points earned outside this "window" will not be considered. As the fitness of an agent would be used to determine which agent to evolve, we do not want its past performances, which might be irrelevant now, to influence the evolution process.

An agent's fitness will always be a positive value, and a new agent would start off with some default fitness. To keep track of the agent's performances, each agent will have a list called *fitness_history*, and this will be used to store the fitness of an agent for each generation. Hence, after an agent is awarded some points, it will calculate its new fitness using Equation (1) and insert the value into the fitness history list.

For our application, the agent's task was designed especially to parse information from a Web site called *Hardwarezone.com.*[4] It is a Web site hosted in Singapore, and it displays up-to-date information of various computer products in table form. The task program allows the agent to establish a connection to the Web site and download the HTML document onto a local computer. The program then parses the document and extracts the relevant information for the agent by looking for specific tags within the HTML document. In our application, the program will be able to extract information such as the description of the product, its price, and the name of the shop selling this product.

## Agent's Knowledge

After an agent retrieved some product information, it needs a place to store this information. As mentioned earlier, when an agent is created, it will register itself to a database. Microsoft Access database is used in this application. Within the database, a table will be created in which each agent can store all the information it retrieved. In addition, the agents will also store this data on a global database. The global database will contain all the products retrieved by the agents in the system.

## Product Recommendation

Before recommending a product to the user, the agent should first be able to evaluate which product would best fit the user's requirements. A proposed method is to use some quantifiable attributes, such as cost, performance, etc., to evaluate the products. An example of an evaluating function could be the following equation:

$$product\_value = perf\_weight*performance - cost\_weight*cost \qquad (2)$$

With the attributes used in Equation (2), we try to model the two types of factors that can influence a user's choice. The first attribute (*performance*) represents the performance of the product, while the second attribute (*cost*) represents the cost of the product. It was assumed that the better the product, the higher will be its *performance*, and a better product usually results in a higher *cost*. From Equation (2), it can be seen that a product with a higher *performance* and/or a lower *cost* will result in a higher *product_value*.

The two weights—*perf_weight* and *cost_weight*—represent the weights that the user could give to each attribute. These two parameters are actually used to represent the user's preferences and are incorporated inside the agent. If *perf_weight* has a higher value, it means that the user placed more emphasis on the performance of the product. Likewise, if the user has a higher value for *cost_weight*, it means that the user is more concerned about the cost of the product.

When an agent is created, these two weights will be initialized based on some heuristics and would be used to calculate the value of each product found in the agent's database. The agent will then rank the products according to their values and select the top three products to be presented to the user. The value of *perf_weight* and *cost_weight* will be allowed to change during agent evolution.

## Agent's GUI

Each agent has a simple GUI that shows information, such as the name of the agent, its current status, products recommended, etc. It would also allow the user to change some of the parameters inside the agents.

*Figure 2: Agent GUI*

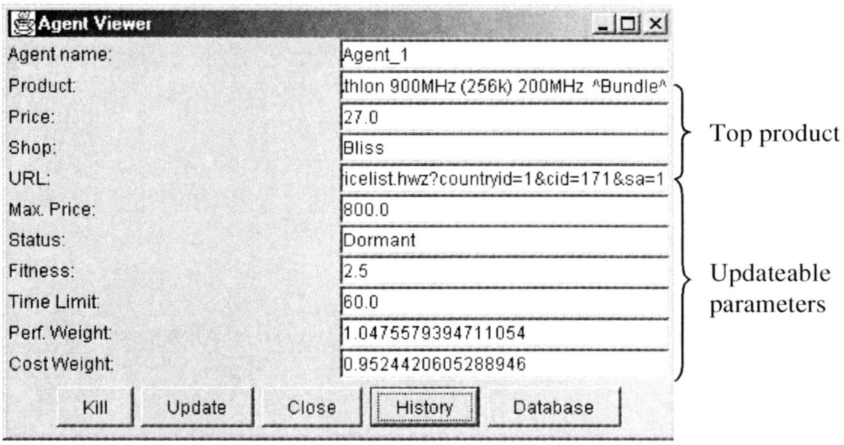

The agent's GUI is implemented as shown in Figure 2. The GUI allows the user to see the top product inside the agent's database and also some of its internal parameters. The user can kill the agent from this GUI by clicking the *kill* button or update some of the agent's parameters by using the *update* button.

## Monitoring Tools

A monitoring tool will be provided for the user that will allow the user to observe and control the behavior of the agents while they search for products on the Internet. This tool will be the main interface between the user and agents.

The user can choose from a list of products provided in the *product ontology* and enter some parameters (e.g., number of agents, etc.) before starting the search. Once all the parameters are entered into the system, the appropriate number of agents will be created to search for the product on the Internet. A screenshot of the implemented system is as shown in Figure 3.

## User Feedback

During user feedback, each agent in the system will select the top three products in its database and add them to a *recommended* list. A sorting function will be implemented to allow the user to sort the list according to his or her preferences. If the user cannot find any product that he or she fancies in this list, the user can look at the *global* list, which contains all the products retrieved by the agents in the system. When the user selects a product that is liked, all the agents in the system will be informed about the user's selection.

*Figure 3: Screenshot of the Monitoring Tool*

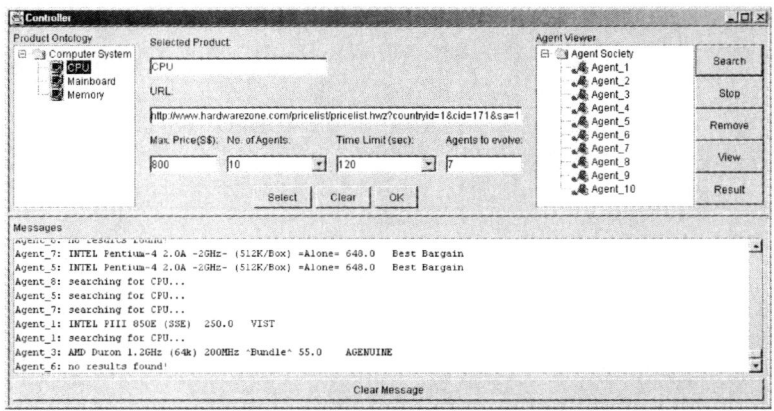

The agents will take note of the product that the user selected and search for that product inside its own database. At this stage, each agent would have already assigned a product value to every product in its database. To determine the amount of points to award to an agent, it will be asked to rank the products in an ascending order according to this value. Hence, products with a higher value will be located at the bottom of the table. The agent will then determine the position of the user-selected product and take note of its row number. The formula to calculate the exact amount of points to give to an agent is as follows:

$$points\ awarded = \frac{row\ number\ of\ user\ selected\ product}{total\ number\ of\ products} \times maximum\ points \quad \textbf{(3)}$$

As an example, assume that shown in Figure 4 is the agent's product list after being sorted in ascending order.

In this example, there are a total of 13 products found inside the agent's database, and the product with the highest product value is located at row 13. This will be the top product inside the agent's database. However, during feedback, the user might have actually chosen the product at row 7 instead. Assuming that five is the maximum amount of points awardable, the amount of points that the agent earns in this case will be as follows:

$$points\ awarded = \frac{7}{13} \times 5 = 2.692$$

*Figure 4: Screenshot of an Agent's Database after Sorting*

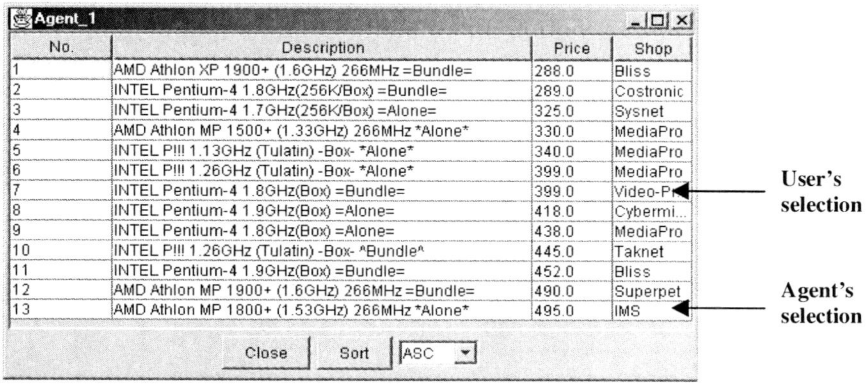

| No. | Description | Price | Shop |
|---|---|---|---|
| 1 | AMD Athlon XP 1900+ (1.6GHz) 266MHz =Bundle= | 288.0 | Bliss |
| 2 | INTEL Pentium-4 1.8GHz(256K/Box) =Bundle= | 289.0 | Costronic |
| 3 | INTEL Pentium-4 1.7GHz(256K/Box) =Alone= | 325.0 | Sysnet |
| 4 | AMD Athlon MP 1500+ (1.33GHz) 266MHz *Alone* | 330.0 | MediaPro |
| 5 | INTEL P!!! 1.13GHz (Tulatin) -Box- *Alone* | 340.0 | MediaPro |
| 6 | INTEL P!!! 1.26GHz (Tulatin) -Box- *Alone* | 399.0 | MediaPro |
| 7 | INTEL Pentium-4 1.8GHz(Box) =Bundle= | 399.0 | Video-P |
| 8 | INTEL Pentium-4 1.9GHz(Box) =Alone= | 418.0 | Cybermi... |
| 9 | INTEL Pentium-4 1.8GHz(Box) =Alone= | 438.0 | MediaPro |
| 10 | INTEL P!!! 1.26GHz (Tulatin) -Box- ^Bundle^ | 445.0 | Taknet |
| 11 | INTEL Pentium-4 1.9GHz(Box) =Bundle= | 452.0 | Bliss |
| 12 | AMD Athlon MP 1900+ (1.6GHz) 266MHz =Bundle= | 490.0 | Superpet |
| 13 | AMD Athlon MP 1800+ (1.53GHz) 266MHz *Alone* | 495.0 | IMS |

## Evolution Process

In conventional genetic algorithms, the agent with a higher fitness will have a higher chance of survival as compared to an agent with a lower fitness. However, in this application, there will be a slight variation in the algorithm. Instead of killing the weaker agents, they will simply copy over all the parameters of the fitter agents. Let *Agent 1* be the fitter agent and *Agent 2* be the weaker one. Shown in Figure 5 is what happens between the agents during evolution.

However, the parameters inherited by a weaker agent in this evolution process might not be optimal. Therefore, the weaker agent will try to adjust its newly acquired *perf_weight* and *cost_weight* to better reflect the user's requirements. First, it will use the newly acquired parameters to reevaluate all the products found inside its new database. Then the agent will select the best product based on these new parameters. If it is the same as the user-selected product, no further changes will be required, but some small and random mutations in the parameters will be allowed.

However, that will not usually be the case. In this case, the agent will compare the *performance* and *cost* attributes of the products that are selected by the user and the agent. Let *p1* and *p2* denote the *performance* of the products selected by the user and agent, respectively. Also let *c1* and *c2* denote the *cost* of the products selected by the user and the agent. Four possible scenarios will have to be considered, and they are as follows:

1.  *p1 > p2* and *c1 > c2*

    The user selected a product that has a much better performance but is more expensive than what the agent suggested. The agent can deduce that

*Figure 5: Evolution Process*

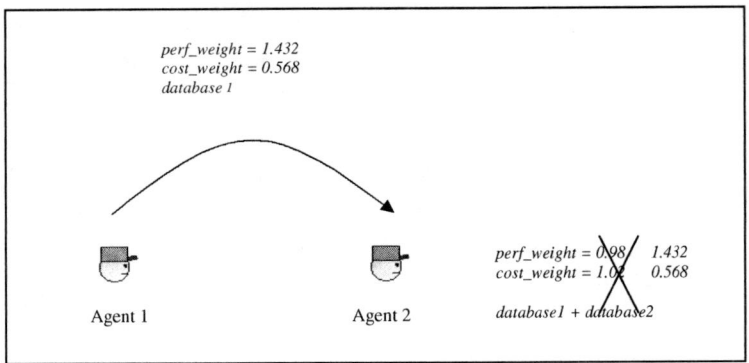

the user places more emphasis on the performance rather than on the cost of the product. Therefore, it will increase its *perf_weight* and reduce its *cost_weight*.

2. $p1 < p2$ and $c1 < c2$

The user selected a product that is of lower performance but is cheaper than what the agent suggested. The agent can deduce that the user places more emphasis on the cost rather than on the performance of the product. Therefore, it will reduce its *perf_weight* and increase its *cost_weight*.

3. $p1 < p2$ and $c1 > c2$

The user selected a product that is of lower performance and is more expensive than what the agent suggested. The agent will be confused over such a selection and will prompt the user if it should carry on the evaluation. If the user still wants the agent to carry on, it will either reduce its *perf_weight* or *cost_weight*. This might happen when the user placed some form of intangible attributes/values on the product that are not present inside the agent's evaluation function.

4. $p1 > p2$ and $c1 < c2$

The user selected a product that is of higher performance and is cheaper than what the agent suggested. This scenario will not arise during evolution. Looking back at Equation (2), a product with a higher performance and/or a cheaper product will result in a higher value being assigned to that product. Because using $p1$ and $c1$ will definitely result in a higher product value as compared to $p2$ and $c2$, this scenario will not happen.

# SYSTEM EVALUATION

To evaluate the performance of the implemented system, some simple experiments were conducted to see if the agents are able to track the user's preference. The computer used in this experiment is a Pentium 4 operating under the Windows ME environment with JDK1.3.0. The system connects to the Internet via a 56.6 kbps modem.

## Product Recommendation

In this experiment, a group of 20 product-brokering agents are instructed to search for CPU on the Internet. Assume that the user aims to get the most

*Figure 6: Recommended List of Products*

| No. | Description | Price | Shop |
|---|---|---|---|
| 1 | AMD Athlon 1.0GHz (2... | 165.0 | IMS Systems |
| 2 | AMD Athlon 1.2GHz (2... | 89.0 | Bliss |
| 3 | AMD Athlon MP 1600+ ... | 385.0 | IMS |
| 4 | AMD Athlon MP 1900+ ... | 490.0 | Superpet |
| 5 | AMD Athlon XP 1600+ (... | 123.0 | io Data |
| 6 | AMD Athlon XP 1800+ (... | 180.0 | io Data |
| 7 | AMD Athlon XP 1900+ (... | 295.0 | io Data |
| 8 | AMD Athlon XP 2000+ (... | 475.0 | Sysnet |
| 9 | AMD Athlon XP 2000+ (... | 309.0 | Video-Pro.com |
| 10 | AMD Duron 800MHz (6... | 88.0 | Laser |
| 11 | INTEL 667A Celeron-S... | 100.0 | BEAM |
| 12 | INTEL 950A Celeron-S... | 65.0 | Video-Pro.com |
| 13 | INTEL Pentium-4 1.8A ... | 379.0 | MediaPro |
| 14 | INTEL PIII 750E (SSE) | 165.0 | HardwarePlace |
| 15 | INTEL XEON 1.7GHz (... | 520.0 | IMS |

Select   Delete   Global List   Agent's list   Close   Sort   ASC

powerful *CPU* and does not care about the price. After instructing the agents to search for the product, the system is allowed to run on its own for about 10 minutes so that the agents can retrieve sufficient products before the user gives feedback. After 10 minutes, the user clicks on the *result* button, and the recommended list is as shown in Figure 6.

From the recommended list, the user selects the current best product at row 13, which happens to be a Pentium 4, 1.8 GHz, as shown in Figure 7.

While the feedback is being made, the system continues to search for products in the background. After making a few similar selections, the agents evolved and reevaluated their list. The new recommended list is now as shown in Figure 8. The list now only shows the best *CPU* retrieved by all the agents.

*Figure 7: User Selection*

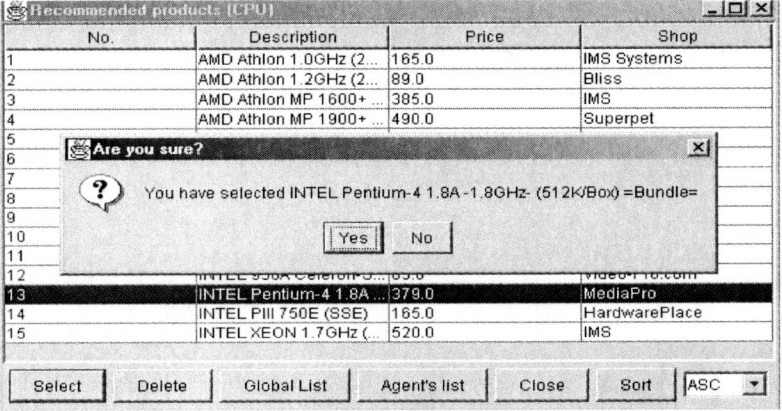

*Figure 8: Recommended List after Feedback*

| No. | Description | Price | Shop |
|---|---|---|---|
| 1 | AMD Athlon XP 2000+ (... | 475.0 | Sysnet |
| 2 | AMD Athlon XP 2000+ (... | 309.0 | Video-Pro.com |
| 3 | INTEL Pentium-4 1.8A ... | 379.0 | MediaPro |
| 4 | INTEL Pentium-4 1.9G... | 452.0 | Bliss |
| 5 | INTEL XEON 1.7GHz (... | 520.0 | IMS |

Select   Delete   Global List   Agent's list   Close   Sort   ASC ▾

When the user is satisfied with what the system learned, the user allows the system to continue searching the Internet for new products on its own. After some time has passed, the agents found an even better-performing *CPU*, and this is reflected in the agent's recommended list, as shown in Figure 9.

## Tracking User Preferences

In this experiment, the objective is to test if the system is able to detect a change in the user's preference and how fast the system will be able to respond to a change. This could be observed by looking at the average fitness of all the agents in the system. The average fitness of all the agents should remain high if the system is able to track and respond to the change effectively.

An initial population of 20 agents is created, and the response of the system is observed by changing the number of agents to evolve in the population.

## Gradual Changes in User Preferences

In the beginning, the user starts by selecting the best *CPU* available. After a few selections, the user will gradually choose cheaper *CPUs*. The experiment

*Figure 9: New Products Recommended by the Agents*

| No. | Description | Price | Shop |
|---|---|---|---|
| 1 | AMD Athlon XP 2000+ (1... | 475.0 | Sysnet |
| 2 | AMD Athlon XP 2000+ (1... | 309.0 | Video-Pro.com |
| 3 | INTEL Pentium-4 1.8A -1... | 379.0 | MediaPro |
| 4 | INTEL Pentium-4 1.9GH... | 375.0 | Video-Pro.com |
| 5 | INTEL Pentium-4 1.9GH... | 452.0 | Bliss |
| 6 | INTEL Pentium-4 2.0A -2... | 545.0 | Global IT Mart |
| 7 | INTEL XEON 1.7GHz (25... | 520.0 | IMS |

New product recommended

Select   Delete   Global List   Agent's list   Close   Sort   ASC ▾

*Figure 10: Tracking Gradual Changes in User Preferences*

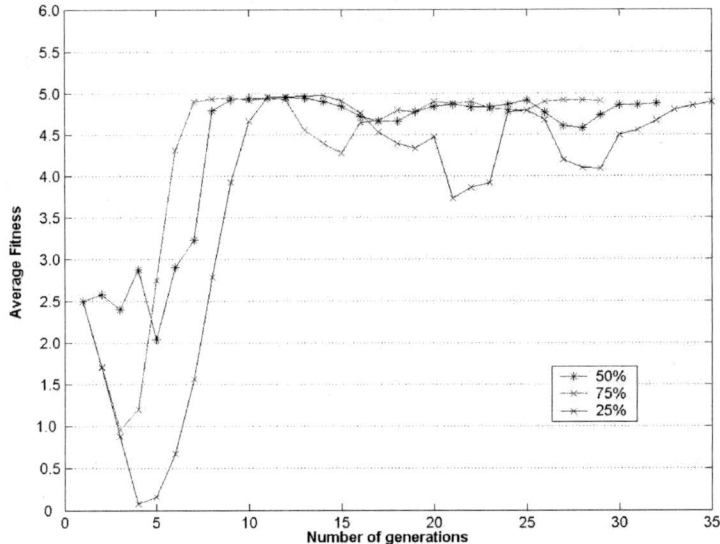

stops after all agents begin to recommend the cheapest CPU available. The average fitness of the agents, when the user gradually changes his preferences, is shown in Figure 10.

The results obtained from this experiment showed that the system is capable of tracking gradual changes in the user's preferences. Although some "dips" are observed during the experiment, the average fitness of the agents in the system remains high while the user changes his or her selection. These dips could happen because some of the agents might not have the products that the user selected in their databases. Therefore, these agents do not receive any points and could significantly "pull down" the average fitness.

## M-Commerce Applications

The proposed design of a product-brokering agent was implemented, using Java on a desktop computer. However, mobile devices such as phones and PDAs tend to have smaller screens, slower processors, and limited memory. Hence, this will pose some serious constraints when we want to transfer the software into these mobile devices. There is also a serious lack of standardization, as these mobile devices use different OS platforms, which makes it difficult for the developer to create a single program that can run on all devices.

After taking these issues into consideration, a possible solution is to use software that is compatible across multiple operating platforms. A good candidate is Java, which has been used to implement the system as mentioned in this chapter. However, the disadvantage of Java as compared to other programming languages, such as C, is its less efficient and slower program execution. Faster processors and more memory are needed to compensate for this. This results in higher cost, and, for wireless applications, shorter battery life. However, this disadvantage has been slowly reduced by the introduction of more efficient JIT (just-in-time) compilers. Recently, the developers of Java also introduced some highly optimized and micro versions of the Java software to cater especially to small devices, such as cellular phones and PDAs.

## Application of Product-Brokering Agent in M-Commerce

A PDA is an ideal device for m-commerce applications. It tends to have a larger screen and a more powerful processor as compared to a cellular phone but is less bulky than a laptop. Making an existing application viewable in any wireless device, a process known as transcoding, is among one of the biggest challenges of m-commerce. In order to fit the screen of the PDA, the GUI implemented in this chapter will have to be scaled down to the appropriate size. A possible solution to fit all these into the PDA screen is to use scrollbars that allow the user to scroll the GUI. A possible screenshot of a PDA with the GUI is as shown in Figure 11.

*Figure 11: Screenshot of a PDA with the Implemented GUI*

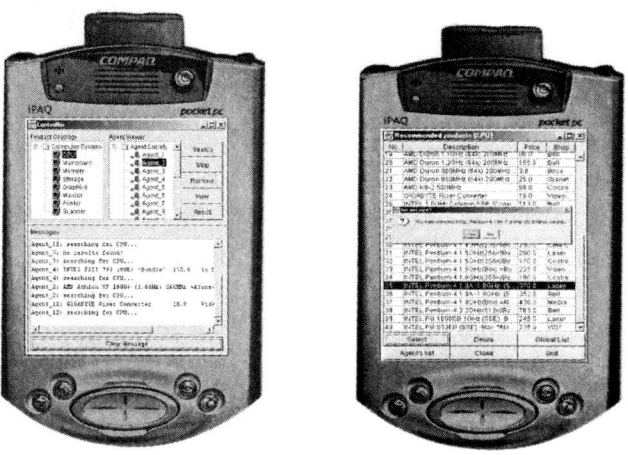

The PDA selected for our application is the Compaq iPAQ Pocket PC H3870. It has one of the largest viewable screens on the market and also has an integrated Bluetooth for wireless links to Bluetooth-enabled cellular phones. This device also supports the Java Virtual Machine, which will allow our software to be integrated easily into the PDA. The specifications of the PDA are as shown in Table 1.

# CONCLUSION AND DISCUSSION

In this chapter, the design and implementation of an intelligent product-brokering agent for m-commerce applications by using genetic algorithms and an evaluation function were presented. A simple prototype was implemented using Java, and the preliminary results obtained from the experiments have been promising.

One of the possible improvements to the current work is to distribute agents in a network instead of hosting entirely in the same computer. For m-commerce applications, this would mean that the agents could now be hosted by a

*Table 1: Specifications for Compaq's iPAQ Pocket PC H3870*

| | |
|---|---|
| **Operating System** | Microsoft Pocket PC 2002 |
| **Processor** | 206 MHz Intel StrongARM 32-bit RISC Processor |
| **Display Type** | Color reflective thin film transistor (TFT) LCD, 64K colors |
| **Resolution / Viewable Image Size** | 240 x 320/<br>2.26 x 3.02 inches |
| **Pixel Pitch** | 0.24 mm |
| **RAM** | 64MB |
| **ROM** | 32MB |
| **Input Method** | Handwriting recognition, soft keyboard, voice record, inking |
| **Wireless Connectivity** | Bluetooth™,<br>Infrared port (115 Kbps) |
| **Dimensions** | 5.3" x 3.3" x .62" |
| **Weight** | 6.7 oz. including battery |

commercial ISP. An m-commerce user would not want to spend huge sums of money on maintaining a wireless connection to the ISP or a phone company. Likewise, it is unrealistic for mobile devices, such as cellular phones and PDAs, to always be "online." Currently, some ISPs provide some forms of storage space for their subscribers to use to store files inside their servers. In extension to this, an ISP could now also offer to host the agents that were created by or for their subscribers with a reasonable fee. These agents could perform their tasks inside these servers and report back to their user when he or she reestablishes another connection with the ISP.

However, allowing agents to be distributed over the network will raise some issues, which the developer should look into before the system could be implemented. Because the agents are distributed, some form of communication protocol and *ACL* (Agent Communication Language) will have to be designed and incorporated into the system.

# ENDNOTES

[1]   http://www.infoworld.com/articles/hn/xml/02/03/22/020322 hnmcommerce.xml.

[2]   http://www.altavista.com.

[3]   Profile-based text summarization.

[4]   http://www.hardwarezone.com.

# REFERENCES

Aylett, R., Brazier, F., Jennings, N., Luck, M., Preist C., & Nwana, H. S. (1998). Agent systems and applications. *The Knowledge Engineering Review*, 13(3), 303-308.

Crabtree, B. & Soltysiak, S. (1998). Identifying and tracking changing interests. *International Journal of Digital Library*, 2(1), 38-53.

Finin, T., Fritzson, R., McKay, D., & McEntire, R. (1994). KQML as an agent communication language. In *Proceedings of the 3rd International Conference on Information and Knowledge Management*.

FIPA Specifications. (2000). Available at: http://www.fipa.org/repository/fipa2000.html.

Maes, P. (1994). Agent that reduce work and information overload. *Communication of the ACM,* 37(7), 31-40.

Morris, S. & Dickinson, P. (2001). *Perfect M-Commerce*. Random House Business.

Nwana, H. S. & Ndumu, D. T. (1996). An introduction to agent technology. *BT Technology Journal,* 14(4), 55-67.

Nwana H. S. & Ndumu, D. T. (1997). Research and development challenges for agent-based systems. *IEE Proceedings on Software Engineering,* 144(1), 2-10.

Pant, G. & Menczer, F. (2002). MySpider: Evolve your own intelligent web crawlers. *Autonomous Agents and Multi-Agent Systems,* 5(2), 221-229.

Sheth, B. & Maes, P. (1993). Evolving agents for personalized information filtering. In *Proceedings of the Ninth Conference on Artificial Intelligence for Applications* (pp. 345-352).

Soltysiak, S. & Crabtree, B. (1998a). Automatic learning of user profiles—Towards the personalisation of agent services. *BT Technology Journal,* 16(3), 110-117.

Soltysiak, S. & Crabtree, B. (1998b). Knowing me, knowing you: Practical issues in the personalization of agent technology. In *Proceedings of the International Conference on the Practical Applications of Agents and Multi-Agent Systems* (pp. 467-484).

# SECTION III:

# WIRELESS COMMUNICATIONS AND MOBILE COMPUTING INFRASTRUCTURE CONSIDERATIONS

Chapter VIII

# Directions in Wireless Telecommunications: Analytical and Operational Pathfinders

John H. Nugent, University of Dallas, USA

## ABSTRACT

*Presented in this chapter are high-level analytical and operational tools and models that assist the wireless telecommunications professional in understanding the telecommunications' market characteristics, life cycles, trends, directions, limits, and drivers. These tools demonstrate that wireless communications will follow the same life cycle characteristics as wire-line communications. Tools are also presented that provide important and timely insight for gaining competitive advantage based upon early detection of critical inflection points.*

## INTRODUCTION

As in all industries, in order to win in a market, it is important to know as much as possible about that market and have at one's disposal tools that will provide insight and competitive advantage when properly, collectively, consistently, and timely applied. In this regard, presented in this chapter is a series of powerful but easy to use and understand analytical and operational tools that deliver insight and competitive advantage to the wireless telecommunications professional. It should be stated, moreover, that as with all good tools, the tools and models as presented herein transition across industry lines and are not limited to the wireless telecommunications industry alone.

These tools and models will be discussed in this chapter.[1]

# MARKET DRIVERS

The concept of market drivers is fundamental to understanding, developing, and prosecuting a winning business case. A driver may be defined as a market fundamental that is basic to meeting a buyer's criteria, need, or desire to have a product or service. A driver, however, constitutes the factor that will move a potential buyer into becoming an actual buyer. This understanding may be best exemplified by one example from AT&T, and another, by looking at an overall model relative to telecommunications equipment manufacturers.

In the AT&T Wireless case, at the time a wholly owned subsidiary of AT&T, Dan Hesse, a brilliant market strategist and at the time president of AT&T Wireless (he is presently CEO of Terabeam), with his team, determined that the fundamental and all-important market drivers in the wireless narrowband communication business in the United States were as listed in Table 1.

Hesse was right. These were the key market drivers in the U.S. narrowband wireless market at the time needed to take this market from primarily a business user market, to one encompassing the nonbusiness user as well. AT&T's narrowband wireless customer base exploded. In fact, most other major narrowband wireless providers were forced to follow AT&T's lead in offering such "One-Rate" pricing programs. These programs led to a major growth spurt in U.S. narrowband wireless subscribers, taking the number of users from approximately 30 million in the mid-1990s to more than 100 million by the end

*Table 1: Mobile Narrowband Wireless Market Drivers*

1. A simple rate plan—one price all the time.
2. No domestic long-distance charges—all calls are local.
3. No time of day or roaming complications or charges.
4. A fixed number of minutes for a fixed price that is deemed "fair."
5. Subsidized handset purchases under term period contracts.

Source: Hilliard Consulting Group, Inc., 2001.

of 2000. Clearly, Hesse and his team's understanding and definition of the "right" drivers for the narrowband wireless marketplace were important to quickly gaining competitive advantage and growing market share.

Conversely, in an important industry example, we see that most industry experts and investment analysts missed predicting the downturn in the telecommunications sector beginning in late 2000 and continuing into 2001. Yet, an understanding of the market fundamentals (drivers) clearly indicated not only that the market would decline but also approximately when.

For instance, from 1996 to 2000/2001, there were *five primary concurrent market drivers* that pushed Telecom equipment capital expenditures rates to approximately 26% cumulative annual growth rate (CAGR) over this period (see Table 2).[2]

1.    **Y2K:** Many telecommunications carriers and other companies had equipment that had hard-coded dates embedded in the equipment's software. It was felt that such equipment would fail or malfunction with the beginning of the year 2000. This was because years ago, when memory was at a premium, programmers coded years by only the last two digits. Hence, it was known that systems would be confused as to whether the year was 1900 or 2000 when we reached January 1, 2000. Many entities believed it was more practical to replace aging equipment than attempt to remediate

*Table 2: The Five Concurrent Telecom Drivers, 1996-2001*

| Period | Driver | Demand Status | Reason |
|--------|--------|---------------|--------|
| 1996–1999 | Y2K equipment replacement | Satisfied | Y2K passed |
| 1996–2000 | Telecom Act of 1996 | Mostly mitigated | Many new entrants failed or consolidated—flawed business cases |
| 1997–present | Wireless digital one-rate plans | Mostly satisfied; markets such as wireless cap out at approximately 50% of the population or 70% of the adult population | Largest demand met, growing from approximately 30 million to 100+ million users—growth slowing significantly: will accelerate with the introduction of broadband |
| 1995–present | Internet usage | Mostly satisfied Growth will accelerate when broadband is more universally available | Largest demand met, growth slowing |
| 1997–present | Circuit to packet | Continuous—the availability of broadband wireless will restart growth here | Transition continues but was slowed by CLEC failure/debacle |

*Source: Hilliard Consulting Group, Inc., 2001.*

it. This replacement of older equipment, which began in some earnest in 1996, was basically completed by December 31, 1999. Hence, this driver was satiated by early 2000.

2. **Deregulation:** Telecom Act of 1996—In 1996 Congress enacted The Telecommunications Act of 1996. This Act created a more deregulated environment that encouraged many new entities to enter the service provider market. Many of these entities, mostly Competitive Local Exchange Carriers (CLECs) and Internet Service Providers (ISPs), raised significant levels of debt and equity funding that they used to buy infrastructure equipment. Unfortunately, most had flawed business cases, resulting in significant financial losses and business failures. In fact, in 2000 alone, 225 CLECs ceased to exist. Further, the market did not fully comprehend how large, entrenched competitors compete—by administrative delay, litigation, and regulatory appeals. These entrenched competitors understood that cash position and flow are king. They had it and the fledging new market entrants did not, so delay through complex administrative matters, legal challenges, and regulatory appeals would likely consume the new entrants' cash positions, as well as the new entrants' abilities to raise more equity or debt because of missed revenue milestones. This competitive practice of entrenched companies was well known and should have served as a red flag to participants in this market and the analysts that covered it.

3. **Wireless:** During this same time, AT&T Wireless introduced the "Digital One Rate Plan." The introduction of this plan drove narrowband wireless utilization of not only Customer Premise Equipment (CPE—handsets, accessories) to high levels, but also drove the requirements for increased levels of network infrastructure to carry the rapidly expanding levels of traffic. Today, narrowband wireless growth slowed significantly, relative to its recent past, and probably will not accelerate until the introduction of GPS-enabled devices and true broadband capabilities. Hence, we see that many equipment manufacturers and wireless service providers did not fully comprehend the dynamics of their market drivers or properly discern approximate levels of market saturation.

4. **Internet:** In the United States, Internet usage started growing in 1995 and continues today. But lately, growth slowed, and it too will likely not accelerate again in the United States until broadband connectivity (the FCC defines broadband as 200 KBPS in both directions) is more universally available. In fact, the timing of this growth is easily seen if we ask ourselves how many of us had an e-mail address in 1995, versus how

*Figure 1: The Telecommunications Market—A Return to Normalcy*

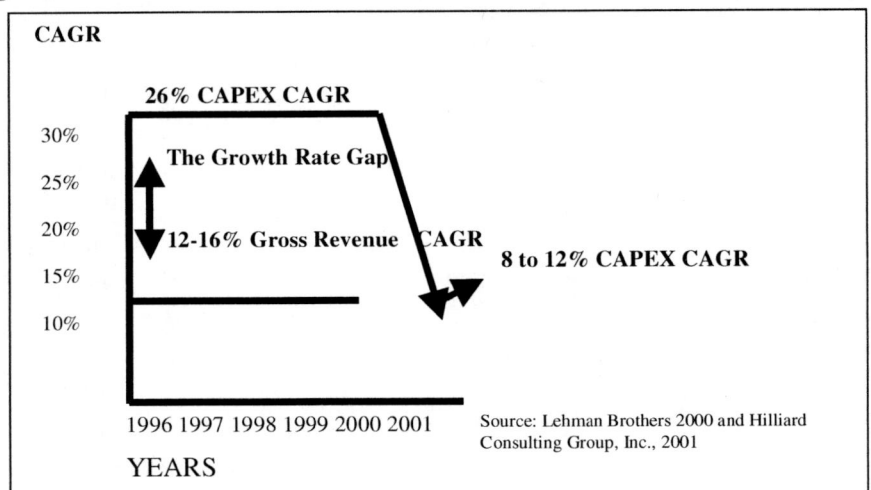

*Source: Lehman Brothers (2000) and Hilliard Consulting Group, Inc. (2001).*

many of us today have one or more e-mail addresses. But this market too is somewhat satiated with narrowband services and needs broadband access to grow significantly again.

5.   **Circuit to packet:** The transition from a circuit to a packet environment will continue, but growth will be slower than would otherwise have been the case had the CLEC failures not taken place and had broadband been more universally available. The impetus for this transition is that packet networks are inherently more efficient in resource utilization than circuit networks. Moreover, as true broadband access becomes more widely available, this transition will again accelerate.

Hence, the main drivers that took place concurrently in the 1996-2001 period created a voracious demand for telecom equipment that has been satiated or mitigated to a large degree (Figure 1).

## Lessons Learned

As we have seen, understanding the right market drivers permitted one company, AT&T Wireless, to grow rapidly and capture market share, while a lack of understanding of market drivers and their linear constraints caused many telecom equipment manufacturers to misperceive the size, breadth, and duration of the telecom equipment and service markets.

In fact, merely plotting the telecom capital expenditures (CAPEX) in 1995 and 1996 relative to gross revenue growth for that same period would have indicated the existing discontinuity between the CAPEX CAGR of 26% and the gross revenue CAGR of 12% to 16% overall for the industry. That is, as early as 1996, we should have seen that an impending correction would take place. And an understanding of the market drivers would have signaled approximately when the correction would take place.

## The Past is Prologue

Now that we examined the recent market past and saw the importance of market drivers, it is important to see if there are additional tools coupled with a view of future drivers that can assist us in discerning the future with more clarity. Here, several tools in particular will assist us in this analysis. The first is the State, Gap, and Trend Analysis.

# STATE GAP AND TREND ANALYSIS

An important analytical tool is the State, Gap, and Trend Analysis (SG&TA).[3] The SG&TA permits us to conceptually define where we are today and project to where we think the market will be at some distant point in the

*Figure 2: State, Gap and Trend Analysis*

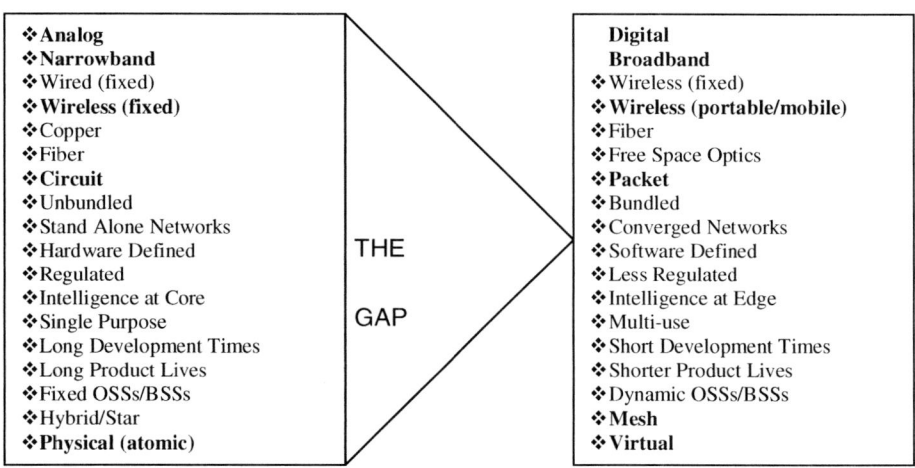

future. With this understanding, we can then discern likely drivers and plot a road map from the present to the future. An example of the telecommunications service provider industry SG&TA is shown in Figure 2.

# THESE SIGNIFICANT TRENDS FAVOR WIRELESS OVER LAND LINES

What this high-level analysis indicates is that the world of telecom service providers and telecom equipment manufacturers is going to change dramatically. Clearly, defining where we are today is not so difficult, and seeing the future also is not so difficult. The hard part is filling in "The Gap" with discrete time lines of where we have to be, by when, based on an understanding of the states of technology and market drivers. Such an analysis may be performed for any industry with an understanding that the model needs to be continually updated. But one thing is abundantly clear—**the telecommunications world is moving to broadband wireless.**

This trend may be seen by examining the projected relative mix shifts and market penetration levels taking place in the market today (Tables 3 and 4, Figure 3).

As can be seen from these tables and figure, there is projected to be a significant move from land-line to wireless, and voice to data communications as a percent of the overall market. Clearly, voice communications is not going to become curtailed; rather, information in the tables indicates that data will grow dramatically in proportion to voice and thus constitute a larger percentage of the mix. However, wireless will make significant inroads into land-line communications, and this trend has already begun, as seen in Figure 3, from Solomon Smith Barney. Moreover, if the Regional Bell Operating Companies

*Table 3: Relative Percent Service Mix Shift*

| Year ⟍ Service | 2002 | 2005 | 20?? |
|---|---|---|---|
| Landline | 80% | 50% | 10% |
| Wireless | 20% | 50% | 90% |

*Source: Hilliard Consulting Group (2001).*

*Table 4: Telecommunications Mode Percent Mix Shift*

| Year / Service | 1995 | 2020 |
|---|---|---|
| Voice | 90% | 10% |
| Data | 10% | 90% |

*Source: Hilliard Consulting Group (2001)*

*Figure 3: Hanging up on the Bells—Technology Elasticity and Service Substitution*

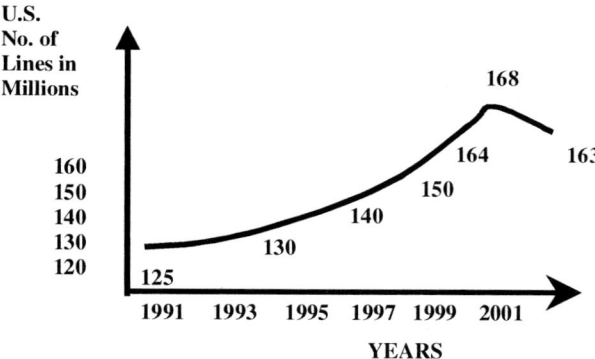

*Source: Solomon Smith Barney Research* Wall Street Journal *4-18-2002, p. B5*

(RBOCs) are successful in having the Federal Communications Commission (FCC) roll back the competitive gains made under the 1996 Telecom Act (UNE-P and TELRIC), the effect of which was to foster more competitive prices, wireless providers should be in an excellent position to offer a wider array of services at very competitive prices relative to land-line solutions, as the RBOCs will likely increase rates once certain competitive elements are removed. So, just as the "Digital One Rate" wireless calling plans first began to penetrate the long-distance land-line market, they too are now making inroads in the local land-line markets, and this trend is likely to continue. Further buttressing this trend is the designation of Western Wireless to "Eligible Carrier Status" in a number of western states, thereby making Western Wireless able to qualify for "Universal Service Funds" for its wireless network services.[4]

# DISRUPTIVE TECHNOLOGIES AND SCHUMPETER'S CREATIVE DESTRUCTION MODEL

Just as a number of jurisdictions awarded Western Wireless "Eligible Carrier Status," signaling improved competitive operating conditions for wireless providers versus land-line carriers, the FCC and the National Telecommunications and Information Administration (NTIA) have both publicly exclaimed in numerous speeches, press statements, and written documents their dissatisfaction with the speed with which traditional land-line carriers are deploying broadband capability to "last mile customers."[5] To this end, the FCC, in 2002, is ardently conducting tests of an extremely disruptive wireless technology—ULTRA WIDEBAND RADIO (UWB).[6] This technology is literally so robust that Intel Corporation is working on a 500 MBPS UWB chip.[7] That data rate is the approximate equivalent of 11 T3s coming to a laptop or other device wirelessly. Other wireless technologies are also pushing the broadband envelope, as shown in Table 5.[8]

This competitive technology landscape is somewhat particular to the United States and several other countries. That is, in the United States, when

*Table 5: Wireless Technology Comparative Table*

| Year | 1974 | 1988 | 2001 | 2003 | 2005 | 2002 | 2002 |
|---|---|---|---|---|---|---|---|
| **Stage** | 1G | 2G | 2.5G | 2.75G | 3G | 4G | 5G |
| Frequency | RF | RF | RF | RF | RF | RF | RF and FSO |
| **Standard** | AMPS | GSM/TDMA/ CDMA | GPRS/ 1XRTT | EDGE/ 1XRTTDO | WCDMA/ CDMA2000 | iBurst UWB MeshNetworks Spitfire Navini IP Wireless | Terabeam AirFiber UWB |
| **Bandwidth** | Narrow | Narrow | Narrow | Narrow | Broad | Broad | Broad |
| **Circuit/packet** | Circuit | Circuit | Packet | Packet | Packet | Packet | Packet |
| **Analog/digital** | Analog | Digital | Digital | Digital | Digital | Digital | Digital |
| **Speed: 4G and 5G definitions are not yet agreed upon** | 9.6 KBPS | 9.6–14 KBPS | 20–144 KBPS | 60–384 KBPS | 384–2000 KBPS | 2000–20,000 (Japan defined) 5000–100,000 KBPS (U.S. defined) | >100,000 KBPS |

one obtains a wireless license from the FCC, the FCC stipulates frequencies, bandwidth, channels, and power authority. That is, in the United States, regulators decoupled technology from the license itself. This is at odds with most other non-U.S. wireless regulators, where the license authority also specifies the technology, i.e., 3G licenses in Europe require W-CDMA technology. This fundamental state in the United States drives Schumpeter's **"creative destruction"** model.[9] That is, no technology is sacrosanct, and as soon as something newer and better comes along, U.S. businesses have been quick to change and envelop the new—without regulatory encumbrances. Intel championed Schumpeter's concept of "creative destruction" by constantly introducing newer and better chip solutions before their existing offerings have run their respective full lives.[10] Such competitive actions permitted Intel to strategically manage prices, moving back and forth from what Michael Porter refers to as "pricing by differentiation" and "pricing to cost."[11] And, Texas Instruments just announced development of a single-chip wireless communications device by 2005.[12] Additionally, we can see from figures that product development and technology adoption rates are taking place in shorter time frames, thereby increasing the chance for miscalculations.

Further defining the competitive landscape and "time-to-market" imperatives, a survey of wireless system and product design engineers indicated a significant compression of development cycle times.[13] This survey yielded the data in Table 6.

Charting these results emphasizes how dramatically professional opinions have changed concerning development times and "time-to-market" imperatives. As we can see, opinions have virtually flip-flopped in just two years

*Table 6: Development Cycle Time Survey Results*

| Perceived Development Time | 1999 | 1997 |
|---|---|---|
| 1 to 3 months | 16.1% | 7.0% |
| 4 to 6 months | 29.6 | 18.5 |
| 7 to 12 months | 27.8 | 26.0 |
| 1 to 2 years | 24.3 | 33.9 |
| >2 years | 2.2 | 14.5 |

*Source: Wireless Design. Adopted by J. Nugent, 1999.*

regarding development cycle time. Complementing cycle time reductions, we also see newer, disruptive technologies generally being adopted at faster rates.

# DISRUPTIVE TECHNOLOGY

From Figure 4, we see that newer disruptive technologies are being adopted in shorter intervals.[14]

Contrasting the adoption rate for the Internet seen in Figure 4, in Table 7, we see that wireless narrowband communications took significantly longer to reach 50 million users.

This lag in wireless narrowband adoption rates was principally due to the relatively high fees carriers charged end users, limiting this market to only business users for a significant period of time. However, from 1997 to 2001, after the introduction of the "Digital One Rate" plans, subscriber growth exploded from 55 million subscribers to over 120 million. This adoption rate for narrowband wireless should indicate to the carriers that the significant driver in

*Figure 4: Technology Adoption Rates*

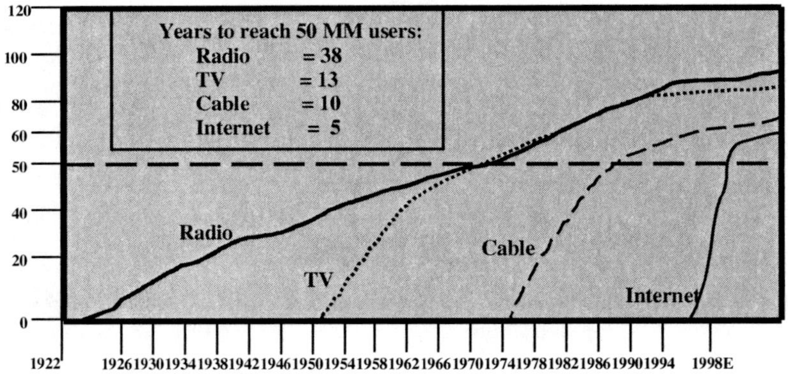

*Source: Lehman Brothers, 1998.*

*Table 7: Wireless Narrowband Adoption Rates*

| Year | 1984 | 1986 | 1990 | 1994 | 1995 | 1996 | **1997** | 1998 | 1999 | 2000 | **2001** | 2002 |
|---|---|---|---|---|---|---|---|---|---|---|---|---|
| **Subscribers** | 0.085 | 0.265 | 5.3 | 24.2 | 33.8 | 44 | **55** | 69 | 86 | 109 | **127** | 144 |

*Sources: Statistical Abstract of the U.S., ITU, CTIA.*

*Table 8: Important Worldwide Trends and Indicators*

| Category | 1990 | 1991 | 1992 | 1993 | 1994 | 1995 | 1996 | 1997 | 1998 | 1999 | 2000 | 2002 Est. |
|---|---|---|---|---|---|---|---|---|---|---|---|---|
| **Main telephone lines (millions)** | 520 | 546 | 574 | 606 | 645 | 692 | 740 | 794 | 848 | 906 | 970 | 1115 |
| **Mobile wireless subscribers (millions)** | 11 | 16 | 23 | 34 | 55 | 91 | 145 | 214 | 319 | 472 | 650 | 1000 |
| **Personal computers (millions)** | 120 | 130 | 150 | 170 | 190 | 230 | 260 | 320 | 370 | 430 | 500 | 670 |
| **Internet users (millions)** | 2.6 | 4.4 | 6.9 | 9.4 | 16 | 34 | 54 | 90 | 149 | 230 | 311 | 500 |

*Source: International Telecommunications Union, 2000 (www.itu.org).*

this market might also pertain to the broadband wireless market. Clearly with computer and Internet usage growing as rapidly as it is, a properly priced mobile/portable wireless access medium would likely be appealing to many subscribers (see also Table 8).

# TOP DOWN, BOTTOMS-UP RECONCILIATION MODEL

Most industry analysts and Wall Street bankers, as well as large corporate financial professionals, are well versed in creating top-down analyses. That is, they are used to undertaking "aggregate" market studies and valuations based on third-party-study house reports of market sizes as well as developing market values based on P/E multiples, EBITDA multiples, NPV/DCF models, or some other aggregate metric. The problem with many third-party market studies is that we already know what they will present before we ever purchase the study. That is, we will see a gross market demand slope over the study period increasing at approximately a 45-60° slope.

The period under study will change from one report to another, as will the metrics on the vertical scale, but the slope is almost always the same (Figure 5).

Hence, we must rationalize aggregate estimates and third-party market studies with a "bottoms-up" analysis.

*Figure 5: Typical Third-Party Market Study Slopes*

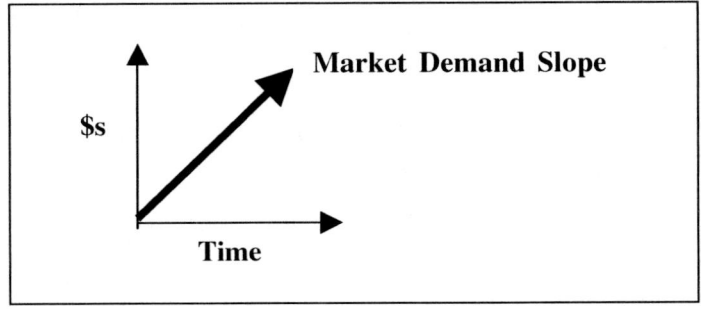

*Source: Hilliard Consulting Group, Inc.(2001).*

# A BOTTOMS-UP APPROACH

A bottoms-up approach mandates that we look at the investment required to reach the market we are after. Here, we can view some investments by AT&T. After a disappointing attempt to capture "last mile customers" with an "almost broadband" service called "Project Angel," AT&T determined that it had to compete with the Incumbent LECs (ILECs) via cable—that is, via a physical connection versus a virtual one. This choice also proved imprudent from an investment perspective, despite what the "Wall Street" pundits were claiming at the time, and serves to illustrate the bottoms-up reconciliation approach. Let us look at an estimate of AT&T's investment and market capture requirements (Table 9).

A market-acceptable rate of return might be taken to be 10%. This being the case, AT&T would have to earn $13 billion in net income annually, beginning in year one, to return an acceptable rate of return. Moreover, AT&T's investment is basically in the current period, while revenue streams would be in future periods; hence, net income streams would have to be

*Table 9: Bottoms-Up Approach—Return and Cost Estimates*

| Cable company acquisitions estimate | $110 billion |
|---|---|
| Two-way broadband upgrades estimate | 15 billion |
| Power solution cost estimate | 5 billion |
| Total infrastructure cost estimate | $130 billion |

discounted by the appropriate present value discount for the respective annual period. Nevertheless, if we accept a general estimate of $13 billion in required net income per year to simplify the calculations, we may ask how many customers AT&T would need to generate this level of net income.

Assuming AT&T's sales staff could generate revenue per customer of $125 per month, or $1500 per annum, for bundled services with a net income after tax per subscriber of 12%, or $180, we can quickly see that AT&T would need 72+ million subscribers ($13 billion divided by $180; well in excess of FCC market limits). Moreover, there are only approximately 280 million people in the United States, represented by approximately 115.9 million U.S. households.[15] Add to this mix a significant number of large well-funded competitors, and we see there was little chance for AT&T to capture sufficient market share to justify such an investment. Typically, market leaders capture just over 30% of any given market. As an aside, studies at General Electric also indicated that to have influence in a market, a player needs 13% to 15% market share.

Here is a wireless case in point. In early 1999, Vodafone PLC announced its acquisition of Airtouch Communications for approximately $62 billion. This was reported to the marketplace as an acquisition at a "good" value, as Vodafone only paid approximately $262 per PoP, Airtouch having licenses that covered 236 million PoPs. However, drilling down, we would have noted that Airtouch only had 1998 revenues of $7.2 billion, and net income for that same year of $560 million. Moreover, its customer base was comprised of 5.4 million wireless phone subscribers and 4.1 million paging customers.

Looking at Airtouch's $560 million net income after tax in 1998, versus what was paid in January 1999, namely $62 billion, we see less than a 1% annual return—far below what could be earned on a risk-free Certificate of Deposit or U.S. Treasury Bill/Note at the time. Additionally, as the investment was up-front, and net income streams have to be present valued over time, we see an investment that appears to return significantly less than risk-free market-acceptable rates of return. Moreover, if we look at what was paid per actual customer versus per PoP, we see that Vodafone paid $6526 per existing customer but only $262 per PoP.

Such excesses in the wireless valuation arena continued into 2000 and 2001, with a number of European countries selling 3G wireless licenses for over $130 billion—just for the licenses! Estimates are as high as another $130 billion for European 3G infrastructure. With an estimated sunk cost of $260 billion (licenses and infrastructure), it is hard to imagine how carriers could ever earn a market-acceptable rate of return on this large of an investment base relative to the anticipated revenue and net income flows.

Undertaking a bottoms-up/top-down analysis, we can look at two of these countries, the United Kingdom and Germany. (For this example, we will treat them as one.) Asking ourselves how many customers they need to justify an investment of approximately $70 billion each, for a total of $140 billion (approximately $35 billion each for licenses, with a similar amount for infrastructure), for a 10% return on the investment (ROI), we can discern the following:

> *Germany in 2000 had an estimated population of 83 million and the United Kingdom 59.5 million. A return of $14 billion annually would be required for a 10% ROI.*

Assuming subscribers will pay an average of $70 per month or $840 per annum, with a net of 10% or $84 per subscriber, we see that Germany and the United Kingdom would require approximately 167 million subscribers ($14B/ $84 per subscriber). This requirement is more than the estimated total population of both countries. Carriers in neither country are likely to achieve market-acceptable rates of return based on such costs and population bases.

Looking at United Nations' population figures, Europe's population is estimated to decline between 2000 and 2050, and Maslov's Laws of Hierarchical Needs are at work here, too.[16] That is, people will not spend on mobile communications until other basic needs are met (food, shelter, security, etc.). So, based on human needs fulfillment and socioeconomic factors, some significant percent of the population will either be too young or not be able to afford 3G services, even if they would like to have them. Hence, a bottoms-up analysis provides fundamental but important information regarding potential investments.

Compounding the 3G dilemma and ultimately affecting values is the introduction of newer, more robust technologies: 4G and 5G. These newer technologies are already entering the market, before 3G has been implemented.

Another reason to temper third-party studies with supplemental assessments is that they often miss important fundamentals when making market prognostications. Here, we only have to look back a few years, when many study houses were projecting stellar growth in the LMDS markets. Well, as we all know now, these markets imploded, and the principal LMDS service providers have filed for bankruptcy protection. So why did these prognosticators miss the mark? Well, there are several reasons.

The first reason is "**technology elasticity**." Technology elasticity is the ability to substitute one solution with numerous substitute technology solutions. That is, many analysts do not understand or adequately address the dynamics of technology development and substitutability and the role that Schumpeter's "Creative Destruction" model plays. As many of the studies often virtually ignore disruptive technologies, it signifies the importance of maintaining a rearview mirror perspective in order to watch for disruptive technologies that may be gaining market acceptance.[17] In fact, a 360° view today is imperative to be, become, and remain a market leader and winner.

# SLOPE ANALYSIS

Another tool that helps provide insight when reviewing business trends and performance comes from a suite of tools coined "Slope Analysis." Slope analysis is based on graphing trends of key variables over time (in the above case, unit price and unit cost), based on actual data, and then extrapolating those slopes to future periods. Such a slope analysis for the IXC market would have indicated, several years before AT&T and MCI proclaimed an understanding of this relationship, that the IXC business model was dead, or at least severely challenged, as currently constructed. That is, the slope analysis would

*Figure 6: Slope Analysis*

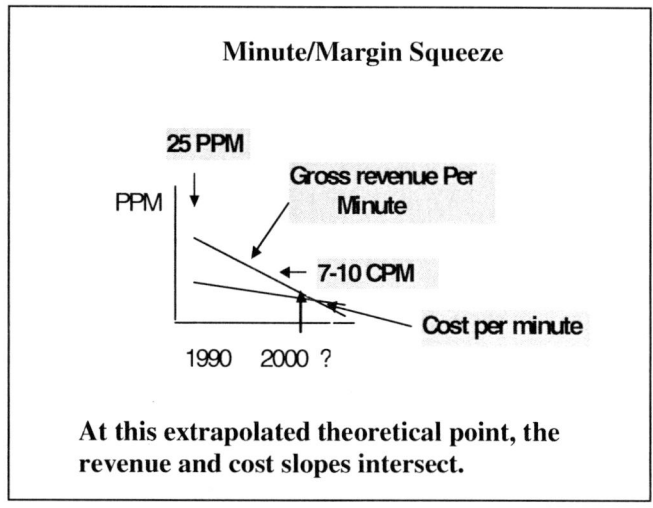

*Source: Hilliard Consulting Goup, Inc. (1995).*

*Figure 7: Average Price/Minute for Mobile Telephone Service*

| Average Price per Minute | $0.45 | $0.53 | $0.58 | $0.57 | $0.56 | $0.54 | **$0.43** | **$0.35** | **$0.28** | **$0.21** | **$0.10–$0.15E** |
|---|---|---|---|---|---|---|---|---|---|---|---|
| Year | 1991 | 1992 | 1993 | 1994 | 1995 | 1996 | 1997 | 1998 | 1999 | 2000 | 2001E |

**Digital One Rates Appear and prices begin dropping precipitously**

*Source: FCC Annual Report on Wireless Industry, June 2001, and Hilliard Consulting Group, Inc. for 2001E.*

have indicated that gross revenue per minute was degrading significantly faster than corresponding costs. When such a relationship comes to be, it indicates through extrapolation that at some point in the probable not too distant future, gross revenue will not provide the margins required to cover costs, let alone make a profit.

The wireless narrowband market may be trending in a similar fashion to the IXC market relative to gross revenue per minute price degradation, as seen in Figure 7.

What is not known at this time is what the wireless carriers are experiencing on a cost per minute basis.

Nevertheless, such a trending, even with estimates, may provide a leading indicator that a business case may shortly no longer be sustainable, and that a market exit or fundamental change in the business case may be required.

## Converging/Diverging Gross Margin Analysis

Another powerful form of slope analysis involves what is known as "Converging/Diverging Gross Margin Analysis." Gross margin is simply the difference between net sales and cost of goods (or services) sold. It is a measure of operational efficiency. **Gross margin may be the single most important metric in any business.**

In a converging/diverging gross margin analysis, we try to determine if the gross margin is increasing as a percent, or decreasing as a percent, of net revenue over time. If we determine that gross margins are increasing as a percent of net revenue over time **(converging = a good thing)**, this indicates that each successive sale is more profitable operationally than those that preceded it. And conversely, if we see a decreasing gross margin as a percent of net revenue over time **(diverging = a bad thing)**, this indicates that each successive sale is operationally less profitable than the sales preceding it (Figure 8).

**Diverging gross margins indicate a serious deteriorating operational condition.** It is important to discern the underlying reasons for a convergence or divergence of gross margins as soon as possible. Understanding why margins are converging may allow one to improve business even further.

Conversely, understanding why gross margins are diverging should highlight fundamental problems as early as possible. These margins may be measured as often as monthly. Seasonal businesses will have to be aware of variations due to the unique natures of their businesses. **Important Note:** Converging gross margins indicate increasing operational efficiencies, as each successive sale is more profitable than the one that preceded it. And conversely, diverging gross margins indicate impairment in the operational efficiency of the enterprise with each successive sale. **Diverging gross margins are a matter of the utmost importance!**

In order to plot these converging or diverging margins, we simply need to capture data from past income statements. A recent 1997 example of this condition was seen when AT&T spun out its wireless business unit. A careful reading of that entity's financials at the time would have indicated that the wireless unit experienced a serious decline in its gross margin—a whooping diverging gross margin in a single year. The importance of this diverging gross

*Figure 8: Converging and Diverging Gross Margins*

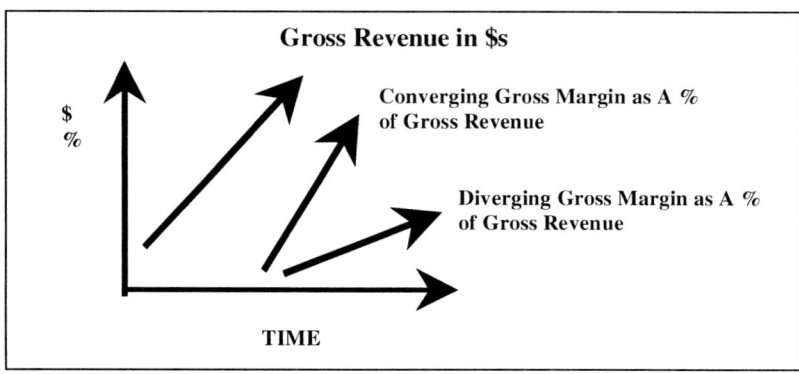

margin was shortly felt by shareholders who bought shares at the IPO price, as soon thereafter, the AT&T Wireless share price dropped precipitously.

Here, we see that the gross margin analysis does not necessarily tell us "why" operations are becoming less efficient, only that they are. Had we drilled down on the AT&T Wireless entity after determining it had a diverging gross margin, we would have seen that this condition arose because of "off-net" or roaming charges paid to other carriers. That is, AT&T marketing and sales organizations introduced a widely accepted calling plan into the market long before AT&T had sufficient infrastructure to complete calls nationally on its own wireless network under its "Digital One Rate Plan." This required AT&T to originate and terminate calls on other carrier's networks at a relatively high cost. This determination of a diverging gross margin in AT&T's case would have further indicated a continuing and relatively large capital requirement in order to build out its network infrastructure in order to bring off-net traffic on-net. And this understanding would have highlighted an impending cash crisis, when viewed in terms of the network buildout required to become operationally cost effective. Hence, a simple converging/diverging gross margin analysis would have led us to better understand the current and continuing issues facing AT&T and AT&T Wireless before others became aware of the issues.

# INFLECTION POINT ANALYSIS

Inflection point analysis (IPA) is extremely helpful in determining when leverage points are in transition.[18] Inflection points are defined as points of major change in any being, one relative to another. In people, inflection points may be identified as marriages, deaths, births, divorces, job changes, home purchases, etc.

Companies 1 and 2 are relatively close in Phase A, but suddenly one gets it, and the other does not. At the inflection point, valuations based upon performance diverge in Phase B.

In business entities, we try to identify inflection points, because they typically indicate one, or more, competitor(s) is/are "getting it," while the others are not, or are not as well. Business is replete with examples of this phenomenon. Pictorially, we may view an inflection point as in Figure 9.

In Table 10, what we see is that one company, relative to others in the pairs, "got it," compared to the others. For instance, in 1998, Apple had a bigger market cap than Microsoft. However, by 1998, Microsoft's market cap

*Figure 9: Inflection Point*

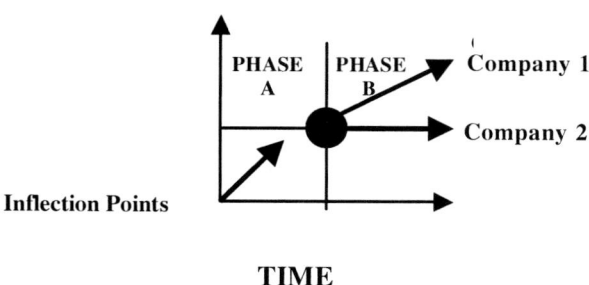

**TIME**

*Table 10: Comparative Company Market Capitalization*

| Group | Company | Year/Market Cap ($s) | Year/Market Cap ($s) |
|---|---|---|---|
| A | Microsoft | **1989/3 billion** | **1998/**220 billion |
|  | **Apple** | **1989/4 billion** | **1998/4 billion** |
|  |  |  |  |
| B | GE | **1983/25 billion** | **1997/**260 billion |
|  | **Westinghouse** | **1983/4 billion** | **1997/17 billion** |
|  |  |  |  |
| C | Cisco | **1993/4 billion** | **1998/**76 billion |
|  | **Bay Networks** | **1993/2 billion** | **1998/6 billion** |
|  |  |  |  |
| D | Nike | **1990/4 billion** | **1998/**10 billion |
|  | **Reebok** | **1990/3 billion** | **1998/2 billion** |

*Source: Examples of major inflection points were highlighted by Slywotzky et al. (1999), as shown in table form.[19]*

far exceeded Apple's. In fact, Apple's market cap remained at $4 billion. Clearly, one company got it, and one did not. Microsoft wanted its software on everyone's computer; Apple only wanted its software on its own hardware. Such conditions are examples of major inflection points.

However, knowing intellectually what an inflection point is and determining the early onset of such a condition is a whole other matter.

# ALTMAN'S Z-SCORE DISCRIMINANT FUNCTION ALGORITHM

In order to attempt to determine the early onset of an inflection point, we are fortunate to have available to us the pioneering work of Edward I. Altman, a professor at the Stern School of Business at New York University.[20] While Altman was looking for a means to predict bankruptcy, his algorithm also works extremely well in determining inflection points and is the basis of the inflection point analysis discussed herein. That is, inflection points as discussed in this work are only used to highlight changes in business condition (positive or negative) not to predict bankruptcy, as Altman envisioned.

Altman used empirical data and regression analysis in order to formulate an algorithm comprised of fractions to which predetermined weights were applied. Scores above or below certain measures indicated the likelihood one would fall into bankruptcy. Altman's tool was found to have a correlation factor of 95% in predicting companies that would file for bankruptcy some 12 months prior to such actual filing, 72% 24 months prior to such a happening, and 48% 36 months before—highly accurate by most measures.

There are some shortcomings. Altman developed the algorithm presented herein for mid-sized manufacturing companies. However, this algorithm works exceedingly well for capital-intensive, infrastructure-laden enterprises—telecom service providers and telecom equipment manufacturers. But, as Altman based his research on mid-sized companies, Altman's time line to business failure is perturbed by the asset-rich nature of the telecom enterprise. That is, the larger the entity (the more assets it has), the more time it has to fix its problems, as it has assets it can borrow against or sell off. This is the AT&T case we see today, with the break up of AT&T and the spin-out of its wireless and cable entities.

Nevertheless, we are only using Altman's algorithm to sense improvement or degradation in business condition (an inflection point), one period to the next. Hence, in our application of Altman's model, **the degree of change in score is significantly more important than the score itself**.

So, how does Altman's algorithm work?

### *Altman's Z-Score—A Discriminant Function Algorithm*

$$Z = 1.2X1 + 1.4X2 + 3.3X3 + 0.6X4 + 1.0X5$$

where:
$X1$ = working capital divided by total assets;
$X2$ = retained earnings divided by total assets;

$X3$ = earnings before interest and taxes (EBIT) divided by total assets;
$X4$ = market value divided by total debt;
$X5$ = sales divided by total assets; and
$Z$ = overall index of corporate fiscal health.

According to Altman, financially strong small- to mid-sized manufacturing companies have a Z-score above 2.99. Companies in serious trouble have Z-scores below 1.81, and companies with scores in between can go either way.

In this case, we can again look to AT&T to see what was happening with its Z-score in the 1998 and 1999 time period (Table 11). Here, numbers were taken from AT&T's Web site, except for the price per share, which came from a financial site.

*Table 11: AT&T's Abridged Financials (Unaudited)*

| Category | 1999 | 1998 |
|---|---|---|
| Current assets | $14B | $14B |
| Current liabilities | 28B | 15B |
| Working capital | (14B) | (1B) |
| Total assets | 131B | 60B |
| Retained earnings | 9B | 8B |
| EBIT | 10B | 9B |
| Equity at market | 161B | 134B |
| Total debt | 82B | 34B |
| Sales | 62B | 53B |
| Shares outstanding | 3,196,436,757@$50 | 2,630,391,784@$51 |
| Gross margin | 53.2% | 51.5% |

*Table 12: AT&T's Z-Score Calculation*

| Factor | 1999 | Score | | 1998 | Score |
|---|---|---|---|---|---|
| X1 | ($14B/131B) | -0.107 | | (1B)/60B | -0.107 |
| X2 | $9B/131B | 0.069 | | $8B/60B | 0.133 |
| X3 | $10B/131B | 0.076 | | $9B/60B | 0.150 |
| X4 | $161B/82B | 1.960 | | $134B/34B | 3.940 |
| X5 | $62B/131B | 0.473 | | $53B/60B | 0.883 |

Applying Altman's algorithm to the AT&T numbers in Table 11, we see the outcomes shown in Table 12.

## Solution

Using the information gathered, we arrive at the following solutions:

*1999:*

$(1.2 \times -0.107) + (1.4 \times 0.069) + (3.3 \times 0.076) + (0.6 \times 1.96) + (1 \times 0.473)$

**= 1.87Z**

*1998:*

$(1.2 \times -0.017) + (1.4 \times 0.133) + (3.3 \times 0.150) + (0.6 \times 3.940) + (1 \times 0.883)$

**= 3.91Z**

As we can see from the above calculations, AT&T's Z-score declined by over half in a single year. **Clearly, AT&T reached a major inflection point**—a major negative inflection point. In fact, if you calculate AT&T's Z-score for 1997, when AT&T Wireless was still a subsidiary, you will find that the Z-score was higher still than 3.91.

As an investor in or creditor or supplier to AT&T, such an early warning that things were heading south would have been helpful in determining your future actions relative to AT&T. For instance, AT&T primarily only agreed to hold unsecured debt. If you were a creditor of AT&T and calculated such a degradation in Z-scores, you might have wanted to lessen your exposure, or raise fees to better offset risk. If you were a supplier to AT&T, for instance, such as Lucent, Nortel, or Ericsson, as AT&T was a large customer, such analysis and determination of a degrading Z-score might have changed your sales strategy relative to this key account. For instance, one of the companies above, if calculating a Z-score for AT&T, might have wanted to offer a discount for large purchases made early, in order to induce AT&T to advance capital equipment purchases. This strategy might have given one supplier a "first grab" at AT&T cash assets long before others realized (perhaps AT&T itself) that capital expenditures might have to be curtailed in future periods. In a moment, we will see how Altman's Z-score algorithm was modified based on negative changes in gross margin.

*Hence, Z-scores, and adjusted Z-scores, can be used to determine inflection points and gain competitive advantage from both offensive and defensive perspectives.*

*Table 13: Adjusted Altman's Z-Score Based on Gross Margin Divergence*

| Annual Decline in Gross Margin % | Reduction in X$ weighting (0.6) | X4 Adjusted Value (0.6) |
|---|---|---|
| 3 < 10 | 100% | 0 |
| >10<20 | 150% | -0.3 |
| >20<30 | 300% | -1.8 |
| >30<40 | 600% | -3.6 |
| >40 | 1000% | -6.0 |

However, as we do not seek to determine the likelihood of bankruptcy but rather only to determine changes in condition (**inflections points**), it is recommended that an adjusted Altman's Z-score model be used. This adjusted model puts more emphasis on debt as gross margin declines. This is because as an entity becomes less operationally efficient, debt becomes more onerous and increases relative risk for the entity.

Reflected in Table 13 is a suggested adjustment to Altman's Z-score model as gross margins diverge in order to highlight significant inflection points.

As we can see in the AT&T Corp. case examined above, as AT&T's gross margin improved for the periods under examination, there would be no need to adjust the X4 factor. But other cases may require such an adjustment.

# CONCLUSION

The tools presented herein demonstrate that industry trends are predictable to a significant degree, as are enterprise inflection points. Such future predictive indicators and early warning red flags permit a better, more timely, allocation or reallocation of resources relative to changing market conditions. An understanding of the models presented indicates that one may gain competitive advantage in a timely manner, permitting better planning and actions that result in superior competitive performance through the application of the tools presented. For instance, an understanding of the Minute/Margin Squeeze Model would have indicated a change in business case fundamentals was required, or failing that, perhaps the need for IXC market exit by approximately late 1998. And, an analysis of the price per minute pressure on wireless carriers today draws a disturbing parallel to the IXC experience. In particular, an understanding of an industry's State, Gap, and Trend Analyses, market drivers, and disruptive technologies permits a better estimate of the composite makeup, size, limits, duration, and shifts in market directions. In sum, collective, consistent, and proper application of the tools and models presented herein should provide significant insight and, likely, competitive advantage in the telecommunications marketplace. Clearly, had these tools been employed by many of the major players in the telecommunications industry over the past decade, we might see a different competitive landscape today.

# ENDNOTES

[1]   Nugent, J. H. (2002). *Plan to win: Analytical and operational tools— Gaining competitive advantage.* New York:_McGraw-Hill. For a complete array of analytical tools and examples, see this reference in its entirety.

[2]   Nugent, J. H. (2001). Telecom downturn was no surprise. *Dallas Fort Worth TechBiz* (www.dfwtechbiz.com), September 10-18, 22. Provided in this article was the analysis that permitted all to see the rapid decline in the telecommunications sector based upon key market driver satiation or mitigation.

[3]   Hilliard Consulting Group, Inc., McKinney, Texas. Model was first developed by J. Nugent at the Hilliard Consulting Group, Inc. in 1999.

[4]   http://www.wwireless.com.

[5]   http://search .ntia.doc.gov and http://www.fcc.gov.

[6]   See http://www.time-domain for an excellent tutorial on ultrawideband radio.

[7]   http://www.intel.com.

[8]   http://www.fcc.com—the FCC generally defines broadband as 200 KBPS in both directions.

[9]   Hunger, J. D., & Wheelan, T. L. *Strategic management* (7th ed.) (p. 300). New York: Prentice Hall.

[10]   *The New York Times,* Intel Corporation insert, June 4, 2002.

[11]   *The New York Times,* Intel Corporation insert, June 4, 2002, p. 62. For a complete explanation as to how to use Porter's discrete competitive models in conjunction with one another versus as opposite solutions, see Nugent, J. H. (2002). *Plan to win: Analytical and operational tools— Gaining competitive advantage.* New York: McGraw-Hill.

[12]   *The Dallas Morning News,* September 3, 2002, and http://www.ti.com.

[13]   *Wireless System Design*, March 1999, p. 9, as adapted by J. Nugent.

[14]   Christensen, C. (1997). *The innovator's dilemma*. Harvard Business School Press. This book provides excellent insight into the realm of disruptive technologies and demonstrates how they can quickly appear and capture entire markets. Note that the book is misnamed. It is actually not the innovator's dilemma, but everyone else's.

[15]   2000 U.S. Census data: http://quickfacts.census.gov/hunits.

[16]   Maslow, A. (1971). *The farther reaches of human nature*. New York: Viking Press; (1943). A theory of human motivation. *Psychological Review*, 50, 370–396.

[17]   For a great examination of disruptive technologies, see Christensen, C. M. (1997). *The innovator's dilemma: When new technologies cause great firms to fail*. Harvard Business School Press. Please note that the author believes this book is misnamed. It is not the Innovator's Dilemma, but everyone else's.

[18]   Based on the pioneering work of Slywotzky, A. J., Morrison, D. J., Moser, T., Mundt, K. A., & Quella, J. A. (1999). *Profit Patterns*. New York: Random House, in their discussion of "The Polarization Phenomenon."

[19]   Slywotzky, A. J., Morrison, D. J., Moser, T., Mundt, K. A., & Quella, J. A. (1999). *Profit Patterns*. New York: Random House.

[20]   Altman, E. I. (1983). *Corporate financial distress: A complete guide to predicting, avoiding, and dealing with bankruptcy*. New York: John Wiley & Sons. The author does not apply Altman's algorithm in its pure form; rather, he uses decimal representations of Altman's weighting functions.

# REFERENCES

Altman, E. I. (1983). *Corporate Financial Distress: A Complete Guide to Predicting, Avoiding, and Dealing with Bankruptcy*. New York: John Wiley & Sons.

Christensen, C. M. (1997). *The Innovator's Dilemma: When New Technologies Cause Great Firms to Fail*. Harvard, MA: Harvard Business School Press.

*The Dallas Morning News*. (2002, September 3).

Hunger, J. D. & Wheelan, T. L. (n.d.). *Strategic Management* (seventh ed.) (p. 300). New York: Prentice Hall.

Maslow, A. (1943). A theory of human motivation. *Psychological Review*, 50, 370-396.

Maslow, A. (1971). *The Farther Reaches of Human Nature*. New York: Viking Press.

*The New York Times*. (2002). Intel Corporation insert, (June 4), 62.

Nugent, J. H. (2001). Telecom downturn was no surprise. *Dallas Fort Worth TechBiz* (www.dfwtechbiz.com), (September 10-18), 22.

Nugent, J. H. (2003). *Plan to Win: Analytical and Operational Tools— Gaining Competitive Advantage (2nd ed.)*. New York: McGraw-Hill.

Slywotzky, A. J., Morrison, D. J., Moser, T., Mundt, K. A., & Quella, J. A. (1999). *Profit Patterns*. New York: Random House.

U.S. Census. (2000). 2000 U.S. Census data. Available at: http://quickfacts.census.gov/hunits.

*Wireless System Design* (1999). (March), 9.

# ELECTRONIC REFERENCES

http://search.ntia.doc.gov/.
http://www.fcc.gov/.
http://www.intel.com/.
http://www.ti.com.
http://www.time-domain/.
http://www.wwireless.com/.

# Chapter IX

# Wireless Middleware

Ken MacGregor, University of Cape Town, South Africa

Nico de Wet, University of Cape Town, South Africa

Bonnie Lam, University of Cape Town, South Africa

Nadim Yazdani, University of Cape Town, South Africa

## ABSTRACT

*Introduced in this chapter is wireless middleware as a means of writing distributed applications for mobile environments. The concepts of middleware and the additional challenges that arise from wireless communications are introduced, in particular, low bandwidth and unreliability. Then described are the commercial wireless products currently available, with particular emphasis on the manner in which they solve challenges. The authors hope that greater appreciation of the capabilities of wireless middleware will enable future developers of applications for mobile environments to produce more efficient systems and researchers to produce better wireless middleware products.*

# INTRODUCTION

## What is Middleware?

Middleware is not a new concept in distributed computing. It was first developed during the 1990s and has evolved significantly since, with the increase in distributed systems. Due to its changing nature, it is difficult to provide one generally accepted definition and scope for *middleware*. A workshop was held at the International Center for Advanced Internet Research in December 1998 to decide on a general definition of middleware and to identify essential services to be researched and developed. Conclusions from the workshop (Aiken et al., 2000) were that it was agreed that the definition of middleware was dependent on the subjective perspective of those trying to define it. It was accepted that:

> *Application environment users and programmers see everything below the API as middleware. Networking gurus see anything above IP as middleware. Those working on applications, tools, and mechanisms between these two extremes see it as somewhere between TCP and the API...*

Perhaps more generic definitions of middleware are (Emmerich, 2000) as follows:

> *"Middleware is a layer between network operating systems and application components." Middleware "facilitates communication and coordination of distributed components."*

This can be visualized in Figure 1.

Middleware has become widely adopted in the industry to simplify the problem of constructing distributed systems. One of the classic application areas for middleware is enterprise application integration, perhaps resulting from corporate mergers. Very often, the period of integration allowed is so short that building a new system is neither feasible nor cost effective. Second, when components are to be integrated, they may have incompatible hardware and operating system (OS) platforms. To build applications using network OS primitives is too time consuming and expensive. Middleware resolves the heterogeneity between systems and provides higher-level primitives so that application engineers can focus on application requirements.

*Figure 1: Middleware in Distributed System Construction*

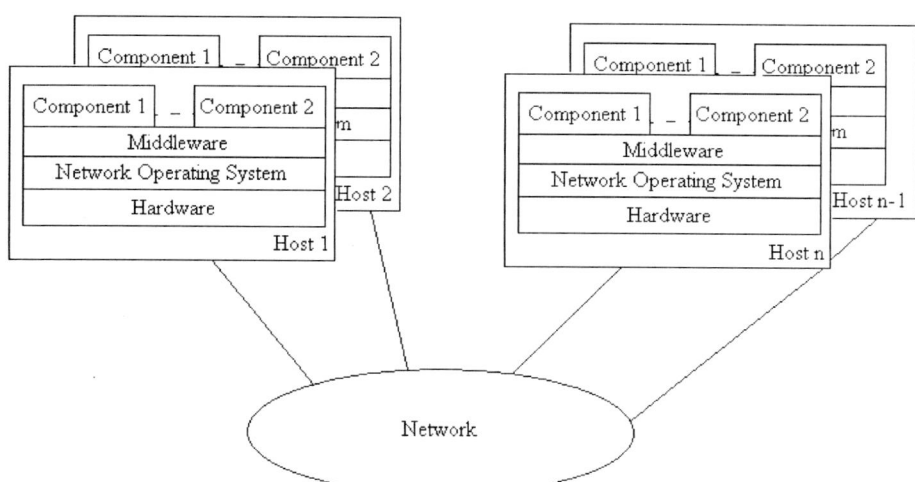

As middleware serves to facilitate communication and coordination between components in a distributed system, it should fulfill the requirements discussed next (Emmerich, 2000).

## Network Communication

Communication between components in a distributed system is achieved by using network protocols. The Transmission Control Protocol (TCP) and User Datagram Protocol (UDP) are examples of protocols used in the transport layer of the Open System Interconnection (OSI) model. Lower layers in the OSI model are managed by the network OS. Application engineers should not need to implement the session and presentation layers. These should be implemented by the middleware. Thus, the middleware should provide the ability to transform complex data structures into a format that can be transmitted using a transport protocol, that is, a sequence of bytes. This transformation is known as *marshalling.*

## Coordination

Synchronization of components communicating with each other is important. This could be achieved in various ways. *Synchronous* communication refers to a component being blocked while waiting for another component to complete execution of a requested service. Synchronization that does not require either of the parties to block on a response from the other party is referred to as *asynchronous* communication.

Sometimes there may be more than two components involved in a service request. This is the case when more than one component is interested in events that occur in another component. This form of communication is then known as *group requests*. Additional support for this type of communication is required to achieve reliable delivery and marshalling of data.

Another coordination issue of activating and deactivating components arises due to the sheer number of components in a distributed system. Middleware should thus provide application programmers to determine the activation policies.

Middleware should support *threading* policies, as a server component may be requested from many client components simultaneously. The server component may be *single-threaded*, in which case, it queues requests and processes them in the order of their arrival, or it can spawn new threads to process each request in a new thread. This policy is then *multithreaded*.

## Reliability

Network protocols have varying degrees of reliability. To ensure that the receiver receives every packet, error detection and correction mechanisms have to be incorporated to handle unreliability. There are four degrees of reliability in communication between components:

- *Best effort.* The service request does not assure execution of the request.
- *At-most-once.* Requests are guaranteed to execute only once.
- *At-least-once.* Requests are guaranteed to be executed, possibly more than once.
- *Exactly once.* Requests are serviced once and only once.

There is a trade-off between performance and reliability. Increase in reliability results in decrease in performance.

For group requests, there are three types of reliability mechanisms:

- *K-reliability.* Indicates that at least *K* components receive the communication.

- *Time-outs.* Specifies the time period after which no delivery of the request is attempted.
- *Totally-ordered.* A request never overtakes a request of a previous group communication.

The above reliability mechanisms are applicable to individual service requests. When one considers reliability for more than one request, transactions are used. Transactions have *ACID* (*atomic, consistency-preserving, isolated, durable*) properties, meaning that a sequence of requests is performed completely, or not at all. Every completed transaction should be consistent. Concurrent transactions are isolated from one another, and once a transaction is committed, it cannot be undone.

Reliability can be further enhanced by replicating components, in other words, making multiple copies of components available on different hosts, so that a replica could service the request should the original component fail.

## Scalability

Middleware should be scalable to accommodate increasing load. In a distributed system, this could be achieved by *load balancing*, whereby the load is distributed across several hosts. Building a scalable distributed system to support changes in the allocation of components to hosts without changing the architecture of the system, or the design and code of any component, is challenging. The International Organization for Standardization (ISO) Open Distributed Processing (ODP) reference model defines, among others, the following two types of transparencies that middleware should support (ISO 7498-1, 1994):

1. *Access transparency.* A component accessing the services of another component is independent of whether the other component is located locally or remotely.
2. *Location transparency.* Components need not know where the other components they interact with are physically situated.

## Heterogeneity

Heterogeneity comes in many aspects for a distributed system. There may be differences in hardware and OS platforms, as well as different programming languages being used for various components. Middleware should be able to resolve this heterogeneity.

# TYPES OF MIDDLEWARE

As middleware has been around for some time, different types have been developed for different applications. The traditional types of middleware have been used for enterprise application integration since the mid 1990s.

## Traditional Middleware

*Messaging Middleware*

Message-oriented middleware (MOM), is specifically built for distributed application development, according to the MOM consortium, which was formed in mid-1993 with the goal of creating standards for messaging middleware.

If a distributed application can tolerate a certain level of time-independent responses, MOM provides the easiest path for creating enterprise and inter-enterprise distributed systems. MOM also helps create nomadic client/server systems that can accumulate outgoing transactions in queues and do a bulk upload when a connection can be established with an office server.

MOM allows general purpose messages to be exchanged in a client/server system using message queues. Applications communicate over networks by putting messages in queues and getting messages from queues (Figure 2). MOM hides all the underlying communications from applications and typically provides a simple high level API to its services commands.

MOM's messaging and queuing allow clients and servers to communicate across a network without being linked by a private, dedicated, logical connection. The clients and servers can run at different times. Messaging does not impose any constraints on an application's structure: If no response is required, none is sent.

*Distributed Transaction-Processing Monitors*

In 1991, the X/Open XTP group published the Distributed Transaction Processing Reference Model (DTPM), which achieved wide acceptance in the industry. The primary purpose of this model is to define the components of a transaction-based system and to locate the interfaces between them. The 1991

*Figure 2: Two-Way Message Queuing*

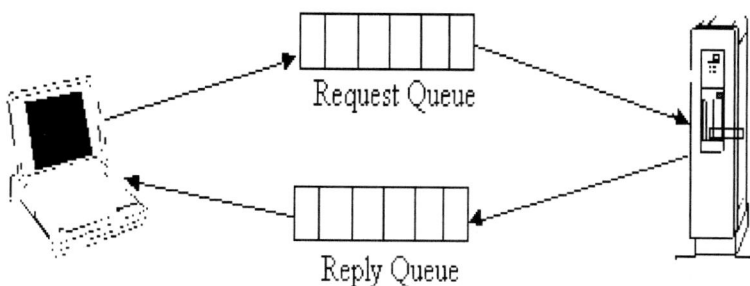

model defined three components: application programs, transaction managers, and resource managers (Figure 3).

A resource manager is any piece of software that manages shared resources, for example, a database manager, a persistent queue, or a transactional file system, and allows the updates to its resources to be externally coordinated via a two-phase commit protocol.

A transaction manager is the component that coordinates and controls the resource managers. The transaction manager and resource manager communicate via X/Open's XA *interface* published in 1991.

TPMs provide a high degree of transactional performance for classic client/server distributed applications, where synchronous communication between the client and server is essential.

*Figure 3: X/Open XTP DTPM Model - 1991*

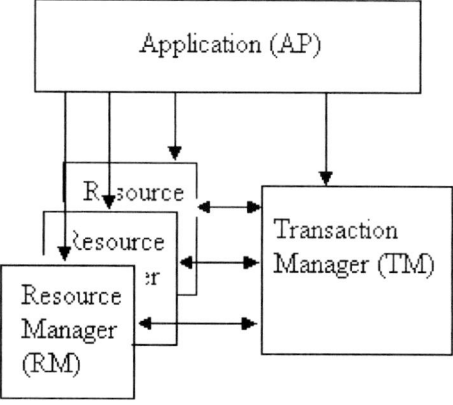

# More Recent Middleware

In recent years, with the advent of Java, use of component-based systems, and prevalence of XML as an integration language, different types of middleware have evolved.

## Component-Based Middleware

Components are blobs of intelligence that can live anywhere on a network. They are packaged as binary components that remote clients can access via method invocations. The language and compiler used to create server objects are totally transparent to clients. Clients do not need to know where the distributed component resides or what OS it executes on. It can be in the same process or on a machine that sits across an intergalactic network. In addition, clients do not need to know how the server object is implemented. For example, a server object could be implemented as a set of C++ classes, or it could be implemented with a million lines of existing COBOL code—the client does not know the difference. What the client needs to know is the interface its server object publishes. This interface serves as a binding contract between clients and servers.

## Component Terminology

The terminology used in components is similar to that used in object-oriented technologies.

A *method* describes one of the behaviors of a component. A collection of methods is called an *interface*, which is the main contract between the software component and the client.

A blob of software can support more than one interface. Thus, a component can be regarded as a collection of interfaces supported by some blob of software. This collection of interfaces is called the *component type*. An *instance* of a component is a particular instantiation of that component. However, we have to consider what is meant by an instance. As the code for all components is shared between all instances of the component, as is the method mapping to the component methods, the only item unique to an instance is the state. Thus, we define an *instance state* to be that unique information that an instance needs to keep. In distributed component systems, it was an obvious abstraction to move the client and component software into different processes that could then be distributed on different machines. To permit the communication between the client process and the component process, a *surrogate* or *proxy* to each process is used. Thus, the client process contains an *instance surrogate* with which it communicates. This passes the message to the *client surrogate* in the component instance component, which passes it to the component instance. Thus, the communication is complete.

### XML Middleware

As one of the main functions of middleware is application integration, it is not surprising that XML was proposed as a middleware standard. One of these proposed middleware products XMIDDLE (Mascolo, Capra, & Emmerich, 2001) has the following use of XML: XML documents can be semantically associated to trees. The data located on the client systems is formatted as XML trees. The applications on the client devices are enabled to manipulate the XML information through the Document Object Model (DOM) (Apparao, 1998) API, which provides primitives for traversing, adding, and deleting nodes to an XML tree. By its use of XML, XMIDDLE can implement the session and presentation layers on top of standard network protocols, such as UDP or TCP.

However, there are disadvantages and advantages in using XML-based middleware. Some of these are as follows:

- Each vendor looks at XML in a different way (a common information exchange mechanism, a database storage format, a format to get metadata under control).
- The exchange of XML information generally requires reformatting twice (coming from the source system moved to the target system) and one mapping (different XML objects).
- XML does not provide a good database format for medium-to-large data sets. There is a large overhead in information access.
- The semantic interpretation of what the data represents is outside the scope of XML.
- The movement of text-based documents can prove extremely inefficient but is always possible.
- XML is good at providing a format for messaging and protocols.
- XML makes sense, because the data is easy to process by the target and the EAI systems.

For these reasons, XML is most commonly considered an Enterprise Application Integration (EAI) tool, rather than middleware, per se.

*Java-Based Middleware*

Java has been in effective use for some seven years now, and with the prolific use of it in application development, it is not surprising that there are many Java-based middleware technologies. Many of these conform to the previous middleware types only in the Java programming language. There is no doubt that despite its somewhat humble start, Java is now mature enough to be of benefit to EAI. It possesses the three necessary criteria of availability, stability, and usability. The Java standards that can be regarded as system interconnectivity are database-oriented (JDBC), interprocess (RMI), message-oriented (JMS) transaction processing (JTS). Of these, the Java Message Server, as it is extensively used in wireless middleware applications and products, will be described in more detail later.

# WIRELESS MIDDLEWARE CHALLENGES AND REQUIREMENTS

Wireless middleware, as a specialized subset of middleware, has come into demand due to the proliferation of wireless networks. Additionally, in the last few years, there has been an upsurge in the development of the mobile devices sector. Personal Digital Assistants (PDAs) and a new generation of cellular phones have the ability to access different types of data. With this, the role of wireless middleware has become increasingly important in providing a reliable channel for data access between mobile devices and servers on a wired network.

Geier (2002) gave wireless middleware the following definition: "Wireless middleware is an intermediate software component that is generally located on a wired network, between the wireless device and the application or data residing on a wired network." The purpose of the middleware is to increase the performance of applications running across the wireless network by serving as a communication facilitator between components that run on wireless and wired devices. Wireless middleware serves as a communication facilitator by addressing the numerous ways in which communication can fail between components in a distributed application. Sources of communication failure, among many others, include a component going offline voluntarily but unexpectedly, a component unexpectedly terminating, a component failing to respond to a request in a reasonable amount of time, and communication being severed in the midst of a request (Sunsted, 1999).

Wireless middleware is presented with many challenges by the very nature of the wireless environment. These challenges are inadequately addressed by standard data communication methods, such as TPM's CORBA, or RMI, which were designed for wired environments. In the wired world, middleware has been used successfully for almost two decades now to allow applications to run on multiple platforms without the need to rewrite them. As such, wireless middleware should support heterogeneous devices and permit applications to be ported from one device to another. However, most traditional middleware products have not been tailored for mobile devices. For example, they require more memory than is available on a PDA and only support communication protocols that were designed for wired networks, notably TCP/IP.

Wireless middleware on mobile devices requires us to deal with challenges not present in traditional fixed-line systems. The most important additional wireless middleware requirements are listed and discussed below.

## Intelligent Restarts

Wireless communication devices tend to lose and regain network connectivity much more often than nonmobile applications. The middleware must be able to cope with intermittent communication links and implement software layers that ensure the delivery of important data between the server and the wireless device. Consequently, the middleware should neither raise communication errors nor lose data when a mobile device loses network coverage. It should incorporate a recovery mechanism that detects when a transmission has been cut, and, when the connection is reestablished, the middleware should resume transmission from the break point instead of at the start of the transmission (Geier, 2002). Intelligent restarts are required in order to enhance the robustness of the distributed system that is built using the middleware. A robust distributed system must detect failures, reconfigure the system so that computations may continue, and recover when a link is repaired.

## Store-and-Forward Messaging

Message queuing is implemented to ensure that users disconnected from the network will receive their messages once the station comes back online.

*Small Footprint*

Mobile applications must be optimized aggressively for small ROM and RAM footprints, as well as for low usage of CPU cycles and battery power. As a result, it is particularly important that the messaging client library, stored in the wireless device, should have a small memory footprint (ROM and RAM).

## Open-Bearer Models

Internet applications only need to support HTTP or TCP/IP. Wireless applications are written to perform on many different networks. A bearer in a wireless network is a transport mechanism of which wireless bearers could be SMS, GPRS, Infrared, Bluetooth, and HTTP. An application written for one bearer typically needs to undergo substantial modifications in order to run on another bearer. The middleware should offer the same set of communication abstractions on top of various wireless bearers. This allows applications to be developed once and operate on top of various bearers. This may, however, lead to the least-rich subset of the possible bearers being implemented.

## Multi-Platform Language Availability

The API should be compatible with languages available on multiple platforms. This allows applications to be adapted for various platforms more easily.

## Security

Security is a key concern in a mobile environment. Security measures must be embedded into wireless middleware applications. As a result, access control, identification, authentication, and end-to-end data encryption are to be provided by the middleware. This dramatically simplifies the development of secure mobile solutions.

## Scalability

As a result of the proliferation of wireless devices, scalability of applications can easily grow to hundreds of thousands of client systems. Ensuring the scalability of applications to this number of clients presents a major problem that must be addressed by wireless middleware. The middleware must not become overloaded when repeatedly sending messages and when encountering increasing congestion resulting from disconnect clients. In addition, it must be able to identify and delete messages, which can never be delivered, perhaps by a wireless device becoming broken or stolen.

## Heterogeneous Software

There are a number of operating systems prevalent in the wireless device market. Because of the memory restrictions placed on these systems, they do not possess all of the features of server operating systems. Thus, it is often difficult to ensure that the middleware can effectively run on all the required hardware devices. Often, features available on one device are not available on another device with a different OS. For example, object serialization, which is a standard method of communication for components between client/server systems, is not available in Windows CE.NET. Thus, programming communications between components where the device is a CE device requires the applications developer to interface with a middleware product, which cannot serialize objects.

### Deployment and Management

Deploying applications to many clients, managing them, supporting them, and ensuring their integrity and support when they are not online, presents another major challenge that must be addressed by middleware.

The basic requirements of wireless middleware can be met by messaging middleware. Messaging middleware serves as a tool for coordinating distributed application components and removes the responsibility for ensuring that messages are delivered reliably and correctly from application components. Messaging products (http://www.softwired-inc.com, http://www.spiritsoft.com) allow distributed application components to communicate and coordinate their activities (via messages) by providing vital services, such as message queuing, message persistence, transactional messaging, guaranteed once-and-only once delivery, and priority delivery (Sunsted, 1999).

As most of the commercial middleware products are based on the Java Messaging System (JMS), we will look at how it can be used as a wireless middleware.

# WIRELESS NETWORKING

As middleware is dependent on the network protocol being used, the effectiveness of the underlying transport protocol must be considered.

## Implementing MOM Using the Java Message Service

The Java Message Service (JMS) developed by Sun Microsystems, Inc. is part of the Java2 Enterprise Edition (J2EE) platform. JMS is an Application Programming Interface (API) for accessing MOM systems in Java Programs. JMS, however, is not only operable within the Java domain, but it can also be integrated with other middleware technologies, such as CORBA and Microsoft.NET.

JMS provides a set of interfaces and good abstraction that defines how wireless middleware can be implemented. For this reason, JMS has become widely accepted in the industry as the API to follow.

The JMS API is not specifically intended for the wireless domain. Traditional JMS was designed to allow Java applications to communicate with existing wire-line MOM systems. As a result, a full JMS implementation is too "fat" for wireless devices, because low power consumption and a smaller memory footprint are required (Spiritsoft, 2002). However, JMS has become the "de facto" standard as an API for wireless middleware.

# THE JMS API

The JMS API allows applications to create, send, receive, and read messages. Messages can arrive asynchronously, meaning the client does not have to specifically request messages in order to receive them. In addition, the programmer can specify different levels of reliability, depending on the type of message transmitted. Unlike traditional low-level network remote procedure call mechanisms, JMS is *loosely coupled*, meaning that the sending and receiving applications do not both have to be available at the same time to enable communication (Sun Microsystems, 2002). In other words, the sender and receiver need not know anything about each other when communicating. This is particularly useful in the wireless domain because of the "sometimes on" characteristics of wireless devices.

# The JMS API Architecture

The following components are present in a JMS implementation:

- *JMS provider.* This component provides administrative and control features. It is the system that implements JMS interfaces. The J2EE 1.3 platform includes a JMS Provider (Sun Microsystems, 2002).
- *JMS clients.* These are components or programs, written in Java, that are the producers and consumers of messages. In the wireless domain, clients may include cellular phones and PDAs.
- *Messages.* These are the objects transmitted between JMS clients.
- *Administered objects.* These are configured by the administrator and include destinations and connection factories. Together, the administered objects form the Java Naming and Directory Interface (JNDI) API namespace. For one client to establish a logical connection with another through the JMS provider, it needs to perform a lookup of administered objects, using a standard directory service.

# The JMS Programming Model

The JMS programming model can be summarized in Figure 4.

- A **Connection Factory** is an administered object that a client uses to create a connection with the JMS Provider. When the JMS program first loads up, the client performs a JNDI lookup of a particular connection factory. A connection factory is either an instance of a QueueConnectionFactory (for point-to-point) or TopicConnectionFactory (for publish/subscribe) interface. Once a connection factory is created, the client specifies the destination to which messages are sent or from which they are received.
- A **Destination** is either a queue or a topic.
- A **Connection** is formed by setting up a virtual connection with a JMS provider. This normally consists of an open TCP/IP socket between the client and the provider. On a wireless link, where bandwidth becomes an issue, normal TCP/IP can prove excessively bulky.
- A **Session** is the context for creating message producer objects, message consumer objects, and messages. Arguments used when creating a session determine whether the session is transacted (considered as an atomic unit of work) as well as the message acknowledgment mechanism.

*Figure 4: The JMS Programming Model*

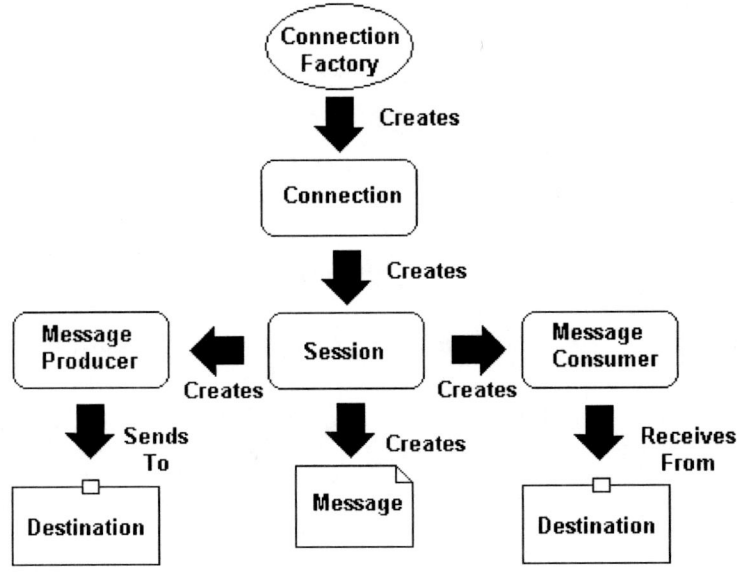

- A **Message Producer** sends messages to a destination.
- A **Message Consumer** receives messages from a destination. The message consumer allows a client to specify interest in receiving messages from a particular destination. The JMS provider is then responsible for delivering messages to the registered consumers. The receive method enables messages to be received synchronously. To receive messages asynchronously, a *message listener* needs to be registered. The message listener's onMessage method, which defines the actions taken on a message, is called whenever a message is delivered. Sometimes, a message consumer may only be interested in a subset of messages for which it is registered to receive. In this case, a *message selector* is registered with the JMS provider. A message selector is simply a string containing an expression and is quite similar in appearance and functionality to an SQL query. **Messages** consist of a header, properties (optional), and a body (optional). Message properties may be predefined or user-defined and can be used, for example, to provide compatibility with existing messaging systems or for setting message selector values. The message body supports numerous data types, such as text files, object messages, and stream messages.

# JMS Reliability Mechanisms

JMS tackles several of the issues inherent in message-oriented middleware in the wireless domain. These include the following:

- Message persistence. What happens if the JMS provider fails or is unavailable during message transfer?
- Message expiration. How does one ensure that unconsumed messages do not cause performance degradation on the system, for example, by filling a particular queue to capacity?
- Durability. Especially in the publish/subscribe domain, what happens to messages published while a subscriber is inactive?

*Message Persistence*

JMS provides two delivery modes for messages. The NON_PERSISTENT mode ensures the lowest overhead but means that messages are not logged to stable storage and are, consequently, lost when a connection to a JMS provider goes down (JMS Specification, 2002). The PERSISTENT mode, on the other hand, is vital for lossy wireless links, where stable connections are unlikely. Messages delivered using the PERSISTENT mode are stored, so that if the link goes down, the JMS provider can reattempt the delivery later. The decision to choose PERSISTENT or NONPERSISTENT delivery is essentially a trade-off between performance and reliability. Clearly, the PERSISTENT mode causes performance degradation due to the overhead involved in storing messages, but it guarantees delivery of messages. The reverse is true for the NONPERSISTENT mode (JMS Specification, 2002).

*Message Expiration*

Each JMS message has an expiration field in its header. A time-to-live value in milliseconds can be specified in this field. The message expiration time is equal to the sum of the current GMT and the time-to-live value. Messages that are not delivered before the expiration time are deleted in order to conserve storage and computing resources (Sun Microsystems, 2002).

*Durability*

The createDurableSubscriber method of the TopicSession class allows a client to have messages retained by the JMS provider while the client is inactive.

The JMS provider stores the messages published to a topic in a persistent queue. When the client reactivates, it is able to retrieve messages published while it was inactive (Sun Microsystems, 2002).

*Other Reliability Mechanisms*

For a message to be considered as having been consumed successfully, it has to be *acknowledged* by the receiving party. However, for performance reasons, it may be inconvenient to acknowledge every message that is received. JMS provides different levels of acknowledgment. When the AUTO_ACKNOWLEDGE value of a particular session is set, all messages belonging to the session are acknowledged. The CLIENT_ACKNOWLEDGE value specifies that messages will be acknowledged on the session level. This means that acknowledging receipt of a message during a session automatically acknowledges receipt of all other messages during that session. The DUPS_OK_ACKNOWLEDGE value allows lazy acknowledgment of messages and is likely to result in duplicate messages being delivered. This value should be set when the client wishes to reduce session overhead but can tolerate duplicate messages arriving (Sun Microsystems, 2002).

Message *priorities* can be set in order to instruct the JMS provider to deliver urgent messages first. Message priorities are specified on a scale from 0 (lowest) to 9 (highest).

JMS sessions can be *transactional*. The first argument to the createSession method specifies whether the session is transacted or not. With transacted sessions, if one operation fails, all changes made during that session are rolled back to a previous committed state. Thus, the operations in the session are seen as an autonomic unit of work (Sun Microsystems, 2002).

# JMS IN COMMERCIAL WIRELESS MIDDLEWARE PRODUCTS

While it can be said that commercial wireless products change rapidly, it is useful to consider the commercial products available at the time of writing. The reader is advised to check the references of these products for additional updated information. JMS provides an acceptable method for communicating in the wireless domain, where unreliable connectivity is expected. The wireless JMS products mentioned below make use of lightweight JMS libraries in order to function effectively.

## SpiritSoft—The SpiritArchitecture

*At the heart of the SpiritArchitecture are two complementary message products—SpiritWave and SpiritLite. These operate with each other (SpiritWave can stand alone) to provide messaging transport across and beyond the enterprise. SpiritWave was the first commercially available JMS implementation...SpiritLite, with its small footprint, low power consumption and minimal load times, runs on any platform and is specifically designed for use by applets (within browsers) and by mobile devices. (Spiritsoft, 2002)*

SpiritLite makes use of SpiritSoft's Java LightWeight Message Service (JLWMS), a stripped-down version of JMS. An application running SpiritLite sends a message to the SpiritWave message server. The SpiritWave server then decides whether the message should be forwarded to another SpiritLite client or transformed to the JMS format for JMS clients. This means that SpiritLite and JMS clients can interoperate seamlessly.

JLWMS is based on the JMS specification but is smaller and less demanding on resources. As an example, JLWMS requires only a single thread to consume messages and makes use of a lightweight wrapper around a JLWMS message payload. JMS messages, on the other hand, impose an overhead of 300 bytes per message even when empty. JLWMS supports point-to-point and publish/subscribe messaging and numerous message types from the JMS specification, including ObjectMessage, TextMessage, MapMessage, and BytesMessage.

Other features of SpiritLite include TCP/IP or HTTP firewall tunneling, serverless messaging between SpiritLite clients, UDP transport instead of TCP when reliability is not an issue, and an open plug-in API that allows new features to be added easily. For example, communication drivers and security tools can be added when required. SpiritLite enables secure, persistent sessions on the Internet with the aid of Secure Socket Layers (SSL).

## Softwired—iBus//Mobile

iBus//Mobile from SoftWired, a Swiss-based company, was first shipped commercially in December 2000 (http://www.softwired-inc.com). The product makes use of wireless JMS (WJMS), a lightweight implementation of JMS. WJMS is capable of implementing both point-to-point and publish/subscribe

*Figure 5: The iBus/Mobile Architecture*

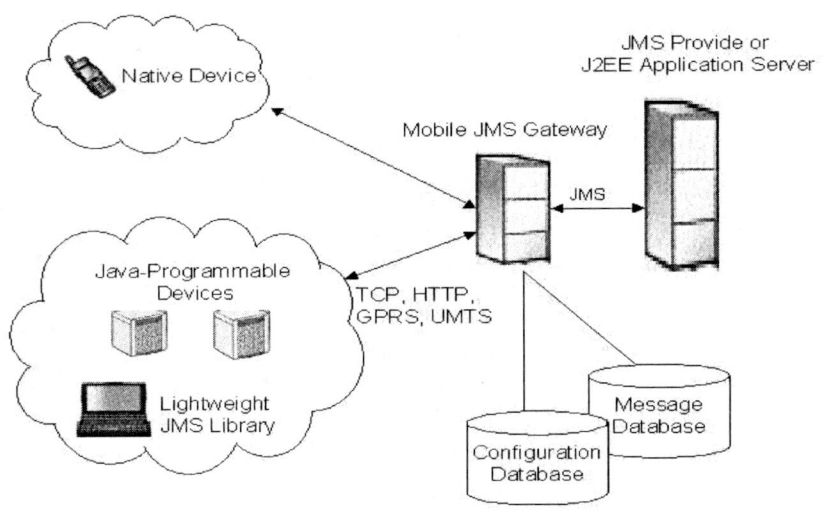

in a Java library of only 70k, and at run-time, an iBus//Mobile application requires as little as 50k heap space (Maffeis2, 2002).

The iBus//Mobile architecture is depicted in Figure 5.

The components of this architecture are a JMS provider, a mobile JMS gateway, and the WJMS client library. The JMS gateway sits in between the clients and JMS provider. To the JMS provider, the gateway appears to be a regular JMS client. From the clients' point of view, however, the gateway acts as a communications hub and message format translator. The gateway guarantees delivery of messages to the receiving party. The WJMS client library is intended for deployment on programmable wireless devices, but as shown above, the iBus//Mobile architecture also caters for nonprogrammable devices, such as pagers and cellular phones. Therefore, it could be said that the client library is optional and depends on the nature of the client device.

*Scalability* is addressed by having several gateways, each supporting hundreds or thousands of mobile devices. The JMS provider would implement load sharing and fault tolerance to cope with the increased demand for resources. Furthermore, two or more JMS providers in a wide area network can be connected using HTTP or SSL. This configuration may appear where a company implements data centers in different regions of a country or even in different countries.

The iBus//Mobile product provides access controls in the form of userID/ password combinations for queues and topics, and encrypts data transmitted between mobile devices and the gateway. iBus//Mobile is able to run on top of most wireless bearers using an appropriate protocol stack. Protocol stacks are instantiated by protocol loaders, and protocol loaders are available for SMS, GPRS, and TCP, among others.

# JMS WIRELESS TRANSPORT PROTOCOL

JMS involves the use of a ConnectionFactory-administered object to create a Connection object. The Connection object is an active connection to a JMS provider. The Connection object, in turn, is used to create the Session and, subsequently, the MessageProducer and MessageConsumer objects.

The crucial point to be considered here is the JMS provider. The JMS specification describes one of the purposes of a Connection object as:

> *It encapsulates an open TCP/IP socket between a client and a Provider's service daemon....Due to the authentication and communication setup done when a Connection is created, a Connection is a relatively heavyweight JMS object. (JMS Specification, 2002)*

Contemporary middleware products, whether message oriented or otherwise, have not been designed for mobile devices (Maffeis2, 2002). The inappropriateness of contemporary middleware design stems partly from the use of communication protocols that were designed for wired networks. The dominant wired network is the Internet, which has the Transmission Control Protocol (TCP) as its transport protocol. Regular TCP, which is TCP designed for wired networks, could be used in the JMS Connection object, however, this may lead to intolerable inefficiencies. Regular TCP as well as the use of regular TCP in networks with wireless links (WLAN, GSM, UMTS) is considered below.

## Regular TCP Features

TCP is a connection-oriented, end-to-end reliable protocol. It facilitates reliable interprocess communication between pairs of processes in host computers attached to distinct but interconnected computer communication networks. TCP uses the services of a less-reliable protocol, such as the Internet Protocol (IP), and provides its own services by means of a variety of facilities. These facilities can be categorized into basic data transfer, reliability, flow control, multiplexing, and connections. The TCP facilities, as described in RFC 793 (University of Southern California, 1981), are briefly discussed below.

### Data Transfer

The TCP has a facility that allows for a continuous stream of octets to be sent in each direction between the users. This is done by packaging variable numbers of octets into segments (such as IP packets) for transmission though the Internet.

### Reliability

TCP recovers from underlying network layer errors. TCP is able to recover from data that is damaged, lost, duplicated, or delivered out of order by the IP. Error-recovery mechanisms used include the use of sequence numbers in packets (to correctly order packets at the receiver) and the requiring of positive acknowledgments (ACK) from the receiving TCP peer. A time-out and retransmission mechanism is used at the sender to deal with ACK packets not arriving. Checksums appended to every packet ensure detection and recovery from damaged packets.

### Flow Control

Flow control is the ability of the receiver to limit (and, hence, control) the amount of data sent by the sender. A sliding window mechanism is used that takes the form of a range (the window) of acceptable sequence numbers beyond the last segment successfully received with every ACK. The window thus represents the set of packets that the sender may send without further permission.

### Multiplexing

Multiplexing occurs when multiple processes within a single host use TCP communication facilities simultaneously. Each host has an IP address and a set of ports. A host IP address and a port form a socket, and two sockets (on different hosts) form a unique connection. A socket may be simultaneously used in multiple connections.

### Connections

A connection encapsulates status information maintained in reliability and flow control mechanisms that are initialized and maintained for each data stream. Status information that is maintained by a connection includes sockets, sequence numbers, and window sizes. Connection establishment between hosts involves the use of a three-way handshake to avoid erroneous initialization of connections (which may occur due to the delayed duplicate problem in unreliable packet-switched networks such as the Internet).

## TCP Congestion Control

RFC2581 (TCP Congestion Control) (Allman, Paxson, & Stevens, 1999) defines the four congestion control algorithms: slow start, congestion avoidance, fast retransmit, and fast recovery. These algorithms are used interchangeably to govern segment transmission based on both the changing network conditions and the response received from the receiver. The algorithms are appropriate for connections that lose traffic primarily because of congestion and buffer exhaustion (Dawkins, Montenegro, Kojo, Magret, & Vaidya, 2001).

Comprehensive coverage of the TCP congestion control algorithms can be found in the RFC, however, they are also briefly discussed here.

- *Slow Start and Congestion Avoidance*. The slow start and congestion avoidance algorithms are used by a TCP sender to control the amount of outstanding data being injected into the network. The slow-start algorithm is used when the TCP sender starts transmission into the network as well as after loss being detected by a retransmission timer. The network conditions are unknown when transmission is started and TCP probes the network to determine the available capacity. Probing the network in this manner avoids congesting the network with an inappropriately large burst of data.

  The slow-start and congestion avoidance algorithms are both used by the sender, however, they are used at different times. Their respective usage depends on the sender's congestion window (cwnd), the receiver's advertised window (rwnd), and the slow-start threshold (ssthresh).

Cwnd is a sender-side limit on the amount of unacknowledged data the sender can transmit into the network. Rwnd is the receiver-side limit on the amount of outstanding data. Ssthresh is used to determine whether the slow-start or congestion avoidance algorithms are used. When cwnd < ssthresh, the slow-start algorithm is used. When cwnd > ssthresh, the congestion avoidance algorithm is used.

### *Fast-Retransmit and Fast-Recovery Algorithms*

The fast-retransmit algorithm is used to detect and repair loss based on incoming duplicate acknowledgment (ACK) packets. The algorithm involves retransmission of what appears to be a missing segment (due to duplicate ACK arrival) without waiting for the retransmission timer to expire. The fast-recovery algorithm controls the transmission of new data immediately after the fast retransmit algorithm sends what appears to be a missing segment, and it remains in control of transmission until a nonduplicate ACK arrives.

## Regular TCP in the Wireless Domain

The performance of regular TCP as described in RFC 793 (Transmission Control Protocol) and RFC 2581 (TCP Congestion Control) is known to be adversely affected by the presence of a wireless link between the sender and receiver. This problem is being and has been researched extensively.

Wireless links have properties that affect TCP performance. Most importantly, they do not provide the degree of reliability that hosts expect. The lack of reliability stems from high uncorrected-error rates (or bit-error rates) of wireless links (especially terrestrial and satellite links) when compared to wired links. Additionally, certain wireless links are subject to intermittent connectivity problems due to handoffs. Handoffs occur in cellular wireless networks such as GSM and involve calls being transferred from between-base transceiver stations in adjacent cells.

The properties of wireless links mentioned have adverse effects on the TCP congestion control algorithms (Dawkins et al., 2001). The root of the problem is that congestion avoidance in the wired Internet is based on the assumption that most packet losses are due to congestion. This assumption is certainly correct in wired links and subnets that have low uncorrected-error rates however, as mentioned, wireless links do not enjoy low uncorrected-error rates.

The result of the incorrect-error rate assumption is poor TCP performance experienced by users. The reason for this observed poor performance is that the TCP connections are spending too much time in congestion-avoidance procedures (such as the slow-start algorithm). Essentially, the TCP sender spends excessive amounts of time waiting for acknowledgments that do not arrive.

The first reaction to packet losses is a drop in the transmission (congestion) window size before retransmitting packets. This is followed by the initiation of congestion control or avoidance mechanisms and backing off of the retransmission timer. The result of these measures is a reduction of the load on intermediate links that results in a reduction of congestion in the network.

The measures mentioned above result in an unnecessary reduction in end-to-end throughput and suboptimal performance. The root of the problem is the excessive amount of time spent avoiding congestion that is triggered by packet losses that result from transmission errors, not congestion. The sender assumes packet loss (say due to congestion-related buffer exhaustion) and thus substantially reduces traffic levels as it probes the network to determine "appropriate" traffic levels.

## Recommendations for Improving TCP Performance in the Wireless Domain

Recommendations (Dawkins et al., 2001; Balakrishnan, Padmanabhan, Seshan, & Katz, 1997) for improving the performance of TCP in wireless and lossy networks can be split into three broad categories: end-to-end protocols in which the sender is aware of the wireless link, link-layer protocols that provide local reliability, and split-connection protocols that break the end-to-end connection at the base station.

The mechanisms for improving TCP performance over wireless links have been comprehensively compared, and methods of improving the performance of TCP over wireless links were presented (Balakrishnan et al., 1997). They state the two fundamentally different approaches to improving TCP performance in the wireless domain.

The first approach involves hiding all noncongestion-related losses from the TCP sender, which consequently requires no changes to existing sender implementations. The reasoning behind such an approach is that the problem is a local one (local to the wireless link) and should thus be solved locally so that the transport layer does not need to be aware of individual link characteristics.

The lossy link is made to appear to be a higher-quality link with a reduced affective bandwidth.

The second approach involves attempting to make the sender aware of wireless links on the path to the receiver. The sender thus knows that certain losses are not due to congestion and can consequently avoid invoking congestion-control algorithms when noncongestion-related losses occur.

### Link-Layer Solutions

Link-layer solutions attempt to hide link-related losses from the TCP sender by using local retransmissions and possibly forwarding error-correcting codes over the wireless link. Local retransmission can use techniques that respond to the specific characteristics of the wireless link, resulting in a significant performance increase. Experiments (Balakrishnan et al., 1997), show a 10% to 30% increase in performance when shielding the TCP sender from duplicate acknowledgments.

### End-to-End Solutions

End-to-end protocols involve the use of two techniques that attempt to make the sender handle losses. The first technique is to use selective acknowl-edgments (SACKs) to allow the sender to recover from multiple-packet losses in a window without having to use a time-out mechanism. The second technique involves the use of an Explicit Loss Notification Mechanism (ELN). It was demonstrated (Balakrishnan et al., 1997) that SACKs and ELN result in significant performance improvements. They show that a simple ELN scheme can improve the end-to-end throughput by a factor of more than two compared to TCP Reno (the de facto TCP standard).

### Split-Connection Solutions

Split-connection approaches completely hide the wireless link from the sender by terminating the TCP connection at the base station. This involves one reliable connection between the sender and the base-station node as well as a second reliable connection between the base-station node and the destination. The second connection can thus use techniques, such as negative or selective acknowledgments as more appropriate alternatives than regular TCP, to perform well (greater throughput) over the wireless link. This technique produced unfavorable results (Balakrishnan et al., 1997), because the sender was found to regularly stall due to time-outs on the wireless connection, resulting in poor end-to-end throughput.

### Header Compression

Improving TCP performance in the wireless domain using header compression is a possibility, regardless of the mechanism used to improve regular TCP performance. The bandwidth of the wireless link will always be limited due to the properties of the physical medium as well as regulatory limits on frequency bands. Limited bandwidth is evident in the circuit-switched GSM data channels that offer 9.6 kbit/s and GPRS extension to GSM that offers up to 170 kbit/s. A consequence of limited bandwidth is the highly desirable process of limiting redundant data transmissions (in the forms of IP, UDP, and TCP headers) in order to efficiently utilize the limited bandwidth. The advent of IPv6, which increases the 20 byte IPv4 header to 40 bytes, serves as further motivation for the implementation of header compression techniques. In fact, the total header size will increase to a total of 84 bytes for a single TCP segment in Mobile IPv6.

Large headers of 50 bytes or more can be reduced in size to four to five bytes, (Degermark, Engan, Nordgen, & Pink, 1997). Their efficient header-compression algorithm takes advantage of the observation that consecutive headers belonging to the same packet stream are either largely identical or seldom change in the life of a packet stream. A rudimentary compression technique, which serves as an introduction to header compression, is mentioned here. However, lossy links, such as wireless links, require more sophisticated measures. It should be noted that header compression has the potential to actually reduce throughput in networks with lossy links, however, Degermark et al. (1997) showed how to avoid such eventualities.

The first step conducted when compressing the headers of the packet stream is for the compressor to send a packet with a *full header*. The full header is a regular header with all fields intact. This initial packet serves the purpose of establishing an association between the nonchanging fields of the header and a *compression identifier (CID)*. The CID is a small unique number also carried by compressed headers. The decompressor stores the initially received full header as a compression state. The second step in the compression and decompression processes is for the received CIDs to be used to look up the appropriate compression state to use for decompression.

The compression technique used means that the CID is sent instead of a full header, thus conserving bandwidth. The compression does not take place for every packet following the initial header. Whenever one of the fields change that are regarded as being mostly unchanged, one needs to send another full header.

## Alternatives to Using TCP

The primary alternative to using TCP is to use UDP with a thin reliability layer that provides the guaranteed message delivery required by JMS. A thin reliability layer is required, because UDP occasionally drops packets and may deliver them out of sequence. UDP packets incorporate checksums that guarantee data integrity, and hence, messages can be reconstructed with a high degree of confidence with respect to message integrity.

RFC 908 and RFC 1151, Reliable Data Protocol (RDP), are designed to provide a reliable data transport service for packet-based applications (applications that send discrete chunks of data such as JMS). RDP can be implemented using UDP and could certainly provide the reliable message delivery required by JMS. RDP can be simple to implement and efficient in the wireless domain:

> *The protocol is intended to be simple to implement but still be efficient in environments where there may be long transmission delays and loss or non sequential delivery of message segments. (Velten, Hinden, & Sax, 1984)*

RDP was developed because TCP has disadvantages when used in certain applications (remote loading and debugging, file transfer, e-mail, transaction processing). RFC 908 lists the general byte-stream transfer of TCP as a disadvantage, due to its complexity, in applications that do not require such a feature.

The argument in favor of using UDP (Bonachea & Hettena, 2000) (with a thin reliability layer such as specified in RFC 908) instead of TCP is that it typically provides the lowest overhead access to the network (it is lightweight) and is widely portable. Additionally, when considering wireless links with their limited bandwidths, UDP becomes attractive due to less header overhead associated with each packet. UDP datagrams contain an 8 byte header, whereas in TCP implementations, one finds 20 to 40 byte headers.

TCP provides a generic protocol, but this generality comes with a price in performance (as mentioned in RFC 908). TCP is a complicated protocol, and software implementations incur significant overhead. Apart from the complexities of the protocol, concerns regarding the scalability of any distributed application are likely to arise. These concerns arise due to significant OS resources, of which there is a finite amount, consumed by a TCP connection.

TCP may be a complex protocol with a significant amount of overhead, yet one can implement a lightweight TCP stack based on the parameters of a particular application or environment, such as wireless links. Apart from the recommendations for improving TCP performance over the wireless link mentioned previously, other techniques have been reported (Frey, 2002). Techniques such as protocol reduction, acknowledgment spoofing, and data compression can be used to improve the efficiency of TCP over the wireless link. On the other hand, performance penalties are likely when using a stream protocol such as TCP for an application, which is message oriented.

## Wireless JMS Provider TCP Solutions

Having examined the alternative transport mechanisms, let us now consider the transport mechanisms used by industrial wireless middleware solutions and the motivation behind the mechanisms used. Our focus is on industrial wireless middleware that implements MOM, such as JMS.

### *Softwired*

Softwired is a wireless JMS solution provider that offers the iBus//Mobile product as their JMS implementation. IBus//Mobile is reported to include an UDP-based reliable messaging protocol that can deliver better performance than TCP:

> *iBus//Mobile provides various features particularly suited for GPRS (and other packet-based bearers). E.g., a UDP-based reliable messaging protocol that can deliver better performance than TCP or HTTP, various transmission parameters that can be tuned (retransmit delays, flow control, transmission windows, etc.), dynamic adaptation to changing bandwidth, reliable messaging in spite of holes in network coverage, data encryption, etc. (Softwired, 2002)*

Softwired does not provide motivation as to why their UDP-based protocol improves on the performance of TCP. Dr. Silvano Maffeis, the inventor of iBus technology, was questioned on the protocol used, and his response was that the information is proprietary. However, he said that the protocol is based on a sliding-window mechanism incorporating both positive and negative acknowledgments. Dr. Maffeis also agreed with our statement that TCP is not optimal for wireless networks.

*Broadbeam*

The Broadbeam Corporation offers Axio as its mobile software platform. The Axio platform includes three components, of which the ExpressQ component is of interest here. ExpressQ is a wireless messaging server offering secure store-and-forward message queueing, real-time communication, and notifications. ExpressQ appears to be remarkably similar to JMS in that it provides a store-and-forward message queuing system with guaranteed message delivery. ExpressQ is reported to provide an optimized wireless transport protocol for reliable and efficient communication (Broadbeam1, 2002).

The following excerpt outlines the optimized wireless transport:

*Optimized Wireless Transport and Compression. ExpressQ was built from the ground up to minimize over-the-air data. ExpressQ's transport layer uses smaller packet headers, uses fewer acknowledgments, produces fewer transmission failures, and retransmits less data over wireless networks than TCP, resulting in significant savings in communication costs. Like TCP/IP, the ExpressQ transport dynamically adapts to changing network conditions (e.g., congestion) to speed up or slow down the rate at which data is transmitted, resulting in fewer failures and errors. The ExpressQ transport automatically compresses and decompresses data to further reduced over-the-air data. ExpressQ's wireless optimized transport and compression result in significantly lower communication costs vs. using transports like TCP/IP. They also result in better application performance, both in terms of transaction times and battery life of mobile devices, and a better user experience. (Broadbeam2, 2002)*

From the above excerpt, one can infer that ExpressQ either uses a lightweight TCP implementation or a reliable data protocol similar to that described in RFC 908. It is clear that regular TCP is not used, and that specific enhancements are included to provide an efficient transport mechanism when traversing the wireless link.

# WIRELESS MIDDLEWARE SERVICES

## Security

Security is essential in a wireless system. As wireless communications cannot be physically secured, the needs of wireless access present a new series of challenges. Wireless access to enterprise systems puts not only the client device, but also the data, well beyond the physical control of the organization. Sniffing of data traffic can be done without any risk of detection, over a much wider range of locations. Furthermore, the client device, in the case of a cell phone or PDA, is even easier to steal than a laptop computer, with the additional loss of security. Compromise of the wireless client thus poses a double threat to data: the remote access to data the device enables, and immediate access to the downloaded data stored within it. The wireless middleware can thus incorporate a key encryption security mechanism as described elsewhere.

## Compression

The low bandwidth in a wireless link requires that the volume of data being transmitted be kept as small as possible. For this reason, data should be compressed from the client to the server and from the server to the client. This desire for efficiency contrasts with the processing requirement on the client side. If the client is a laptop computer, then it can have adequate processing power to decompress the data; however, if the client device is a PDA, the amount of processing power available currently is inadequate to decompress large amounts of data in real time.

As optimal-compression algorithms are data specific, the compression algorithm to be used will vary on the type of data. For example, text compression, picture compression, Java class compression, and XML compression, all use specific, different algorithms. The best the middleware can do is to have a library of common compression algorithms it can use to compress uncompressed data depending on the type. Alternatively, the problem of compression can be regarded as an application design consideration and be left to the application developer to elect whether to compress the data or not. The middleware will then only pass data it is given and not attempt to perform any compression.

# FUTURE TRENDS OF
# WIRELESS MIDDLEWARE

From what we described above, wireless middleware has grown from the traditional middleware background to having a life of its own. The early attempts of stating that standard wired middleware products are "suitable for wireless" have passed, and the wireless middleware products in commercial use today have different requirements, use different transport protocols, and have different concepts of robustness of connection.

In the future, we will see these products enhanced with a rich array of features to handle different application domains. Already, XML-based middleware is produced that uses XML as its internal structure. Middleware incorporating wireless payments as an additional feature is available. In the future, wireless middleware will be tailored to include specialized applications for entertainment, gambling, and synchronization of applications. However, the underlying features will be the basic facilities as described in this chapter.

# REFERENCES

Aiken, B., Strassner, J., Carpenter, B., Foster, I., Lynch, C., Mambretti, J., Moore, R., & Teitelbaum, B. (2000, February). Network policy and services: A report of a workshop on middleware. *Network working group—Request for comments 2768*. Retrieved May, 7, 2003 from the World Wide Web: http://www.faqs.org.

Allman, M., Paxson, V., & Stevens, W. (1999, April). TCP congestion control. *Network working group—Request for comments 2581*. Retrieved May, 7, 2003 from the World Wide Web: http://www.faqs.org.

Apparao, V., et al. (1998, October). *Document object model (DOM) Level 1 specification*. Retrieved May, 7, 2003 from the World Wide Web: http://www.w3.org/TR/1998/REC-DOM-Level-1-19981001.

Balakrishnan, H., Padmanabhan, V. N., Seshan, S., & Katz, R. H. (1997, December). A comparison of mechanisms for improving TCP performance over wireless links.

Bonachea, D. & Hettena, D. (2000). Amudp: Active messages over UDP. Retrieved May 7, 2003, from the World Wide Web: http://www.cs.berkeley.edu/~bonachea/amudp/.

Broadbeam1 Corporation. (2002). *ExpressQ*. Retrieved July 1, 2002 from the World Wide Web: http://www.broadbeam.com/pdf/expressq.pdf.

Broadbeam2 Corporation. (2002). Axio, Broadbeams mobile software platform. Retrieved July 1, 2002, from the World Wide Web: http://
www.broadbeam.com/pdf/axio_white_paper.pdf.

Dawkins, S., Montenegro, G., Kojo, M., Magret, V., & Vaidya, N. (2001,
August). End-to-end performance implications of links with errors.
*Network working group—Request for comments 3155.* Retrieved
May 7, 2003, from the World Wide Web: http://www.faqs.org.

Degermark, M., Engan, M., Nordgen, B., & Pink, S. (1997, October). Low-
loss TCP/IP header compression for wireless networks. *Wireless Networks,* 3(5), 375-387.

Emmerich, W. (2000). Software engineering and middleware: A roadmap.
*Proceedings of the Future of Software Engineering* (pp. 117-129).
Limerick, Ireland.

Frey, A. (2002). New middleware paradigm for wireless network computing.
Retrieved May 7, 2003, from the World Wide Web: http://
www.networkcomputing.com/715/715wswire3.html.

Geier, J. (2002). Wireless network middleware. Retrieved April 2002 from the
World Wide Web: http://www.wireless-nets.com/articles/
whitepaper_middleware.htm.

*IEEE/ACM Transactions on Networking (TON),* 5(6), 756-769.

ISO 7498-1. (1994). Information processing systems—Open systems interconnection—Basic reference model: The basic model. Retrieved May 7,
2003, from the World Wide Web: http://www.iso.org.

JMS Specification. (2002, April). Java Message Service Specification Version 1.1. Retrieved April 2002 from the World Wide Web: http://
java.sun.com/products/jms/index.html.

Maffeis1, S. (2002). *Introducing Wireless JMS.* Retrieved April 2002 from
the World Wide Web: http://www.softwired-inc.com.

Maffeis2, S. (2002). JMS for Mobile Applications and Wireless Communications. Retrieved April 2002 from the World Wide Web: www.softwired-
inc.com/people/maffeis/ articles/softwired/profjms_ch11.pdf p17.

Mascolo, C., Capra, L., & Emmerich, W. (2001, August). An XML based
middleware for peer-to-peer computing. *Proceedings of the IEEE
International Conference on Peer-to-Peer Computing (P2P2001).*
Linkoping, Sweden.

Softwired. (2002). *iBus//Mobile Frequently Asked Questions FAQ.* Retrieved April 2002 from the World Wide Web: http://www.softwired-
inc.com/products/faq/faq-mobile21.html.

SpiritSoft. (2002). Wireless & JMS: Extending the enterprise to real time wireless messaging. Retrieved April 2002 from the World Wide Web: http://www.spiritsoft.com/download_files/select.asp.

Sun Microsystems. (2002). Java Message Service Tutorial. Retrieved May 7, 2003, from the World Wide Web: http://java.sun.com/products/jms/tutorial/1_3_1-fcs/doc/overview.html.

Sundsted, T. (1999). Messaging makes its move. Retrieved April 2002 from the World Wide Web: http://www.javaworld.com/javaworld/jw-02-1999/jw-02-howto.html.

University of Southern California. (1981). Transmission control protocol. *Network working group—Request for comments 793*. Retrieved May 7, 2003, from the World Wide Web: http://www.faqs.org.

Velten, D., Hinden, R., & Sax, J. (1984). Reliable data protocol. *Network working group—Request for comments 908*. Retrieved May 7, 2003, from the World Wide Web: http://www.faqs.org.

**Chapter X**

# Usability Issues and Limitations of Mobile Devices

Suliman Al-Hawamdeh, University of Oklahoma, USA

## ABSTRACT

*Due to the fast development of mobile technologies and wireless communications, more people are using mobile devices. Mobile devices like cellular phones are initially used for data communications, as in speech. Now, mobile devices are not only portable but also can be used to communicate and exchange information as well as gain access to remote services anywhere, anytime. But while mobile devices offer many opportunities for e-commerce applications, conducting e-commerce transactions over mobile devices has its limitations. Limitations include limited memory, limited processing power, different technologies and standards, small keyboards, and small screens. A usability study was carried out to determine the extent to which mobile devices can be used in mobile commerce. Most of the studies showed that while mobile devices are becoming increasingly popular with the younger generation, users still prefer to use desktops for e-commerce transactions. This is mainly due to the limitations of mobile devices and the stability and security of the wireless networks.*

# INTRODUCTION

Mobile devices are getting better every day. Over the past few years, the world has seen an explosion of new devices like cellular phones, Palm Pilot, Pocket PC, and Auto PC. Mobile applications can now be developed to deliver different types of information to users around the world. Different mobile devices support different programming languages, such as WAP, WML, HTML, and Java. WAP is a global specification that allows mobile users to access information and services instantly through wireless devices (http://www.wapforum.org). The NTT equivalent to the WAP protocol is i-mode®. NTT DoCoMo first introduced i-mode in Japan in February 1999, and as of today, it has more than 15 million users (http://www.nttdocomo.com). Unlike i-mode, which is available only from NTT DoCoMO in Japan, WAP is offered by many competing organizations throughout the world.

A major difference between the WAP and i-mode is that WAP runs WAP protocol, while i-mode runs on HTTP protocol. Development in WAP encompasses languages such as WML and WMLScript. In i-mode, the markup language used is cHTML (compact HTML). cHTML requires a cHTML gateway before a site can be accessed by users from their i-mode-compatible mobile phones. Similar to WAP, the underlying network technology does not matter. As such, the underlying network technology can be shared with Internet Web sites. Another application development language differentiate between WAP and i-mode is that while WAP uses cards to display WAP pages, i-mode does not facilitate the use of cards. This difference in concept is due to the different standards that have been adopted by WAP and i-mode. While the use of WMLScript can be difficult in the sense that missing or incorrect tags could crash the whole application, the use of cHTML is friendlier. The i-mode is able to display rich GIF and JPEG images, tables, multiple font types and font sizes, background colors, and style sheets on its browser. This is unlike WAP, which only displays WBMP images to WAP browsers.

WAP is a joint standardization effort for converging Internet and value-added services (VAS) to wireless devices like mobile telephones, pagers, and personal digital assistants (PDAs). Examples of VAS implementation include Short Message Services (SMS), mobile fax and data communication, auto roaming, e-mail updates, voicemail, and mobile Internet. This standardization effort is carried out by the WAP Forum, founded in June 1997 by a group of companies, namely, Ericsson, Motorola, Nokia, and Phone.com (formerly known as Unwired Planet). WAP is a universal open standard accepted by standardized bodies like the Telecommunications Industry Association (TIA) and World Wide Web Consortium (W3C).

WAP started as a result of the increasing demand for messaging and data services known as SMS, supported by the Global System for Mobile communications (GSM). After this, High-Speed Circuit-Switched Data (HSCSD) supported high-speed services for business users in 1999. HSCSD enables the transmission of data over a GSM link at speeds up to 57.6 kbit/s. By using HSCSD, a permanent connection is established between the caller and the calling parties for the exchange of data. Because it is circuit switched, HSCSD is more suited to applications such as videoconferencing and multimedia than "bursty"-type applications such as e-mail, which is more suited to packet-switched data (Furuskar, Naslund, & Olofsson, 1999).

WAP is developing rapidly due to support and demand from end-users who feel that using mobile devices is more personal than using a PC. This is because mobile devices like wireless phones and pocket devices are on all of the time (Compton, 2000). Furthermore, users of these devices can gain access to Internet services anytime and anywhere. People who require these types of services are "on-the-go" most of the time (consultants, managers, and staff who need to travel more often). Due to the development of mobile communications technologies in the 1990s, more people are able to and can afford to use mobile devices. These mobile devices, like cellular phones, are initially used for data communications, as in speech. Now, these mobile devices are not only portable, but they are also becoming the tools that people use to communicate and exchange information as well as gain access to services remotely, anywhere and at any time.

One of the common uses of remote services is electronic commerce (e-commerce). But, conducting e-commerce transactions over mobile devices has its limitations. Issues like limited memory, processing power, lack of standards, small keyboards, and small screens affect the amount and types of data that can be used in e-commerce transactions. Currently, most e-commerce applications available for mobile devices are limited to short transactions in the form of messaging transactions. In this chapter, we investigate the usability of mobile devices in e-commerce. As Singapore is considered one of the highest users of mobile devices in the world, this study explores users' concerns and acceptance of mobile devices. It also compares the use of mobile devices, such as cellular phones, PDAs, and laptops, to the fixed network using desktop computers. Issues covered in the study include sound effect, multimedia effect, user-friendliness of the user interface, navigation, speed of access to usage, users' preferences, convenience of usage, security of usage, and size of screen.

# MOBILE COMMERCE

E-commerce is a term used to describe business conducted using electronic means. Young in 1995 stated that "Electronic Commerce is the buying and selling of goods and services where part, if not all, of commercial transaction occurs over an electronic medium." E-commerce utilizes different types of networks, like VAN, the Internet, intranet, and extranet. VAN is a network where the service provider provides the transport network and services for document interchange, mainly electronic data interchange (EDI) applications. EDI is about doing business and carrying out transactions with trading partners electronically. EDI covers most things that are done using paper-based communication, for example, placing orders with suppliers and carrying out financial transactions. This is why the term "paperless trading" is often used to describe EDI (Cloberg, 1995). The Internet is a global network and does not have a single owner or a centralized administrator or control. Intranets are internal networks that are built in organizations to utilize Internet technologies. In order to provide security between the Internet and intranet, firewalls are utilized.

E-commerce helps to break down traditional barriers of geography, time, and size. Location, time of access, and the size of the organization does not matter anymore on the Internet (Chin & Chua, 2000). Figures from Boston Consulting Group (BCG) show that online stock trading is the Internet's killer application in Asia. It is also the second-largest retail e-commerce category, after hardware and software sales. The number of women online shoppers is set to increase as more are getting connected to the Internet and e-tailers. And, e-retailers are starting to target them. In small countries like Singapore and Hong Kong, online retailers will have tough competition with "bricks-and-mortar" rivals. The reason is that most shoppers find it easy and convenient to go to the shops situated next door, compared to that in bigger countries like Australia and the United States.

According to the e-commerce director for Barclays Stockbrokers, Phillip Bungey, WAP technology will enable information to find the client, as opposed to the other way around, and it will make transactions possible on the move. The company will provide proactive personalized market information to its clients by using WAP technology (www.hp.com). This information is also known as "push" information, where users can personalize the information received. A combination of geographical positioning system (GPS) service and WAP service could provide location-based services. An example of such service is the WAP-based "Restaurant Finder," where information like restaurant phone number, address, menu details, pricing, etc., are gathered and delivered to customers' mobile devices wherever they are. This is made possible by producing a location-based WAP service, combining location with content to the mobile users. The aim is to provide a convenient and easy-to-use WAP service.

Wireless chats and e-mails, multiplayer gaming, friend finder, and preloaded software (that provides personal planner, diary, organizer, alarm clock, and calculator) are available in the United States (http://inf2.pira.co.uk/top002.htm). Besides providing entertainment and personal management software, other service providers, such as CNN Mobile and Reuters Wireless Services, have been providing news, weather, and stock market information to subscribers' phones. With mobile service, users have the power to trade locally and globally. They will also not miss any opportunity to trade. Moreover, the service is promised to be easy and convenient, by merely pushing a few buttons on the mobile phone. Fast-food chains like Kentucky Fried Chicken, Edo Sushi, and Pizza Hut introduced ordering through Internet. Besides being able to order food through the Internet, customers are also able to do it using WAP-enabled gadgets. This is one of Singapore's fast-food portals. A customer is required to log on to the fast-foods.net site on his WAP device to get the service (Oo, 2000). The Internet version of fast-foods.net has received over 3000 orders since it was launched in March 2000.

# LIMITATIONS OF MOBILE DEVICES

While mobile devices make it easier for users to communicate and exchange information on the move, many of the users still get frustrated when they try to carry out e-commerce transactions using their mobile devices. This is largely due to the fact that mobile devices are new, and the technology associated with these devices is still evolving. It will take time before these limitations can be addressed. In this section, we briefly discuss some of these limitations and highlight some of the research and developments taking place in each area:

1. **Limited memory**

   Most mobile devices have limited memory storage. This is set to change in the near future. Samsung Electronics Co., Ltd, which is considered one of the industry's leading suppliers of advanced semiconductor memory technology, announced that it has begun to work on a 2 Gbit NAND Flash memory utilizing the 90 nanometer design. The NAND Flash memory can be set in a 4 Gbyte memory card to store data equivalent to 70 CDs or 40 minutes of DVD-quality video data.

2. **Limited processing power**

   Many of the services provided over the Internet, such as gaming, music, and video, require faster machines and higher memory levels. It is expected that the Intel PXA250 and PXA210 chips will enable mobile phones and handheld computers to deliver music, video, and games at higher speeds and better quality.

3. **Different technologies and standards**

   Until now, vendors competed to provide mobile devices without adhering to any particular standards. This will change in the future, as mobile device manufacturers start to realize that it is in their interests to cooperate and use open standard technology. In May 2002, Nokia and Siemens Information and Communication Mobile announced that the companies agreed on a framework of collaboration to create and drive the implementation of mobile terminal software based on open standards. Such efforts will help to open the door for more players and more collaboration.

4. **Small keyboards and input method**

   Most mobile devices have only a numeric keypad and a small number of cursor keys. This limits their usage and makes them more difficult for users to use to interact with services and navigate through remote Internet sites containing large amounts of content. Given that, keyboard and mouse interaction could be replaced by voice- and pen-based interactions.

5. **Screen size and color**

   While screen resolution continues to improve, and color screens are becoming the norm, screen sizes are likely to remain small, as users prefer small and portable devices. However, this might change when electronic paper becomes available. Electronic paper is a paper-like sheet made up of thousands of microcapsules that are electrically charged to display white or black ink, or any other pair of colors. When this technology is commercialized, it will revolutionize the mobile devices industry, due to the fact that electronic papers can be folded and stored in a small pocket.

6.  **Battery consumption**
    Batteries are important for the mobility and portability of mobile devices. Batteries run out quickly in most handheld devices. This is especially true if the user is on the move and has no time to charge the battery. There is an ongoing effort to reduce consumption of power and increase battery life for mobile devices. Some of these efforts include new battery technologies, like the fuel cell.

7.  **Simplicity of user interface**
    When you are on the move, you cannot afford to read instructions or manuals when you want to perform certain operations. The ability to access applications and data within, literally, a few keystrokes is very important in mobile devices. User interfaces need to be friendly and attractive. Graphical capability is another issue that constrains the development of attractive interfaces on mobile devices.

8.  **Limited bandwidth**
    As the number of mobile users increases, there will be greater demand for shared wireless capacity. There is still a lot of development work to be done to improve the limitations of mobile devices and network bandwidth. New products, such as Wi-Fi5, which is based on high-speed standards and runs in 5 Ghz spectrums and provides up to 54 Mbps, are starting to reach the market. The network bandwidth is expected to improve in the next few years, and the number of users is expected to grow as well.

9.  **Stability and dropouts and connectivity**
    For many users on the move, connectivity is very important. In order to maintain connectivity and minimize dropouts, mobile devices should have the ability to save and minimize the loss of data, while conducting transactions.

10. **Voice recognition**
    Pen-based interaction on a touch-screen could potentially replace the mouse, while voice-based interaction could be used for activation and control of functions like voice dialing. Access control for voice recognition and speed could enhance the usability of mobile devices and open the door for new types of applications.

# PENETRATION OF MOBILE DEVICES

In a 2001 report by Datamonitor, in the United States, titled "U.S. Mobile Devices to 2006, a Land of Opportunities," it was stated that there were 81.7 million mobile phone shipments in the United States by 2001. Multiplying at a compound annual growth of 17%, this figure will reach 183.8 million in 2006. In the meantime, shipments of handheld mobile devices will reach 49.9 million in year 2006. This will represent an increase of 22% from the year 2001. The report forecasted that the U.S. market will replicate Europe's success in wireless communication and growth by twofold in the next five years.

A survey, titled "Are you Ready? The use of mobile data applications in Europe," carried out by Arthur Anderson in 2000, focused on the corporate business market for mobile data and included 749 companies in Europe. In the survey, 78% of the participants viewed mobiles phones as critical or very critical to their business. Convenience and speed were some of the reasons for the popularity of mobile phones. Other benefits included the ability to gain operational efficiency, improve communications within the organization, especially with staff members constantly on the move, and the ability to be contacted from anywhere at anytime. Although more than three quarters of the participants admitted that mobile phones were critical to the survival of their business, the survey unraveled that few companies in Europe had been using applications that run on mobile platforms to deliver services to their customers. That was evident by the fact that only 6% of the respondents used mobile phones for e-commerce transactions.

While many people saw mobile phones as potential business delivery channels with which to improve relationships, the survey results showed that slow speed, limited bandwidth, high cost, etc., were some of the barriers for using mobile devices as a business delivery platform. Most of the respondents said that they were concerned about WAP security issues. The survey also showed that only 9% used mobile phones to access financial information, 11% used mobile phones to access information such as traffic and weather, and 8% used mobile phones to access news. Despite the fact that mobile phones were not used for e-commerce-related activities, it is generally expected that this will change, as the services provided by mobile devices improve and become more affordable and secure. The survey also showed that 24% of the respondents were using PDAs, 21% used WAP phones, while another 44% planned to use WAP-enabled mobile phones.

Vodafone, a mobile phone operation in Britain, working jointly with Ford, a car manufacturer, to bring wireless Internet to cars (Rudkin, 2001). Such wireless Internet service will be made available by pushing buttons on the car's stereo system. The wireless service will supply traffic information and help in the event of emergencies to the drivers of Ford cars, by facilitating GPS technology. In essence, on top of providing traffic information, the system will provide directions to drivers on how to get to a selected destination. It will also trigger emergency services whenever the car airbag is activated. Future services to be provided by Vodafone and Ford will include messaging services, weather information, and stock quotes. These services have been available in Germany since March 1, 2001. Other target implementation locations include Britain and other European markets.

The use of WAP was extended to create an organization's IntraWAP services (Loken, IntraWap—Wireless Offices, 2001). This service is similar to intranet services that were available through a desktop-bound Internet browser. A WAP-enabled phone can be used to access an organization's intranet site outside the office location. Such a service eliminates current problems with an intranet services modem and a notebook computer.

WAP technology was also extended to allow town councils in Europe to extend services to the public (Loken, 2000). A wireless pilot program in Stockholm, which was jointly developed by Telia, a mobile phone operator in Stockholm, and Stockholm Parking Corporation, allows a driver to park at a parking space and calls a specific number to register the parking lot. When it is time to leave, all the driver needs to do is to call the same telephone number, and the parking fee will be added to the customer's telephone bill. Such a wireless service allows consumers to park their cars conveniently by eliminating the need for consumers to queue and the need to have coins or small change to pay parking fees.

The use of WAP and mobile devices in entertainment and gaming businesses has great potential (Rudkin, 2001). In the United States, Digital Avenue, a WAP games developer company, put some of its most popular games onto its mobile portal (http://www.digitalavenue.8m.com). Games including Black-jack for gambling fans, Power Hockey for sports fans, Quiz for intelligent games fans, Monkey Island for adventure game lovers, and many other games applications, such as Trivial Pursuit, have been brought to WAP devices through SMS technology by Motorola and Codeonline (Loken, 2001). Codeonline is a Finnish mobile entertainment company. The games are expected to be available to all Motorola partners and will be available in the United Kingdom from April 2001.

Animated messages were brought to European WAP users by FunMail (http://www.funmail.com). FunMail is the organization that brought full-color animated messages to i-mode users in Japan. With such animated color messages, a WAP user will be able to send nontextual-based messages to another WAP phone user. Hewlett Packard Singapore reported that the company and Intel awarded a grant of $1 million to help five local companies deploy wireless solutions developed at Hewlett Packard's Mobile E Service Bazaar (http://myfsi.hp.com/solutions/esip/wp.pdf). The grant will help various WAP applications, such as mobile connectivity to office and personal e-mail, booking of cinema tickets via wireless devices, and wireless order and inventory services, to be introduced to the Singapore market. The grant, in essence, will help development and implementation of wireless technology to be boosted in Singapore and, in return, help to promote the development of wireless services to allow Singapore-based companies to tap into wireless technology and that consumer base.

Hewlett Packard, in September 2000, announced that it signed an agreement with Singapore Telecom and Lycos Asia to launch a product called HP Wireprint (http://www.hp.com.sg/news/2000-09-27.html). This e-service allows users to retrieve and send information over a secure channel to mobile devices or other output devices like printers and facsimiles. Unlike traditional mobile printing, HP Wireprint does not require the presence of the source file on the mobile devices. Furthermore, the printer drivers do not need to be installed on the mobile devices. From the user's side, all that is needed is user identification and password in order to access the shared resources, such as a network-shared drive on the network in the office. The service will first be deployed in the Asia-Pacific region and will then progressively expand world-wide.

The Singapore government is also interested in adopting some WAP applications to deliver public services. Such applications add additional channels for government to deliver public services and add convenience to public. The Ministry of Defense (MINDEF) extended its IPPT (a physical fitness test) booking service to national servicemen using WAP-enabled phones (http://www.ippt.mindef.gov.sg/ippt/wap/defencetown.wml). MINDEF also allows national servicemen to register their intention to temporarily leave the country through its Going Overseas WAP application.

The stockbroking industry is one area where WAP applications and mobile devices can be used more effectively. In Singapore, two stockbroking firms brought online trading facilities to WAP technology. They are POEMS and Fraser Securities. The WAP site of POEMS allows consumers to trade stocks, unit trust, and account information over WAP-enabled mobile phones (http://www.poems.com.sg). The WAP trading is available only to POEMS customers who subscribe to Mobile One, Singapore Telecom, or Starhub mobile phone operators. There are selected mobile phones that had been certified to be WAP-trading enabled. Services available in WAP trading include the service to allow consumers to place an order to trade in The Singapore Stock Exchange and other linked exchanges. Consumers are also able to view the status of their orders online, through WAP-enabled mobile phones. Consumers can also set alerts so that when a given counter reaches a given price, an alert is sent from the POEMS computer server to the consumer's mobile phone. Other services include account information facility, whereby consumers could check on their shares position, the amount of money that is due to or from POEMS.

Fraser Securities Pte Ltd, another stockbroking firm, provides online share-trading services via a mobile network to share investors in Singapore. Its Web site is currently accessible through Singapore Telecom E-Ideas. E-Ideas is a mobile commerce product from Singapore Telecom, where m-commerce services are grouped into categories for easy access (http://www.fraser direct.com.sg). The trading features provided by Fraser Securities are similar to those provided by POEMS. The difference is only in the mobile phone operator. While POEMS allows customers to access its site through any one of the three mobile phone operators in Singapore, Fraser Securities allows its customers to access its site only through Singapore Telecom.

Banks also start to offer WAP services using mobile devices. DBS Bank customers who possess a WAP mobile phone or handheld devices would be able to use the WAP-based banking services through DBS Wireless Banking (http://www.dbs.com/ebanking/wireless). DBS Wireless Banking services allow customers to check their account balances and perform fund transfers between different accounts. DBS WAP services allow customers to access bank rates (e.g., fixed deposit rates, loan rates, etc.) and request financial news (e.g., news on stocks, shares, and corporate announcements) from the bank. Customers can also use the service to locate DBS or POSBANK offices, branches, and ATMs. For security, the WAP services use WTLS (Wireless Transport Layer Security) to protect a transaction by utilizing encryption technology. The WTLS security module that the bank implemented supports up to 128-bit key strength.

In the telecommunication sectors, the three mobile phone operators in Singapore, namely, Singapore Telecom, Mobile One, and Starhub, have also launched their WAP services. In comparison, the three companies are offering services that are similar in nature to those services offered by telecommunication companies in countries such as Britain and the United States. Singapore

Telecom launched a WAP service available through mobile phones and PDAs called E-Ideas (Intelligent Do It Yourself Electronic Access Services). Singapore Telecom customers who signed up for this service are able to have an E-Ideas menu in their mobile phone and PDA devices. The menu lists a selection of e-commerce sites accessible through their mobile phones. The service allows customers to use their mobile phones to perform online transactions, such as electronic trading and electronic banking, online taxi reservation, and e-mail, any time of the day and anywhere in the world, as long as Singapore Telecom mobile phone roaming service is available. According to Singapore Telecom, there will be more services, such as electronic ticketing, electronic games, etc., planned to enrich Singapore Telecom E-Ideas (Emanuel, 2000). E-Ideas is aimed to benefit people constantly on the move, enabling them to access a wide range of information and services.

Starhub is another telecom player who introduced a personalized WAP service through its innovative product called iPower. iPower allows consumers to access the Internet without using a desktop computer. Unlike SingTel's E-Ideas, iPower allows personalization of the customer's mobile phone. The personalization allows customers to organize their bookmarks to the WAP site. Such personalization can be configured from the Starhub Web site. Starhub also extends its e-mail services to its mobile phone customers. Customers are allocated an e-mail address such as [telephone-number]@starhubmail.com.sg. This e-mail service allows customers to send and receive e-mail messages from their mobile phones. Basic e-mail services, such as managing messages and an address book, are also provided.

Mobile One Pte Ltd (M1), one of the three mobile phone operators in Singapore, launched Mi World, a WAP service available exclusively to M1 customers. Like its competitors, it allows its customers to get access to the latest financial, local, and international news, check weather and traffic conditions, e-mails, games, and banking with OUB Bank. M1 also allows its customers to purchase hampers and cards using their mobile phones. M1 presents WAP share-trading services through POEMS and Keppel Securities to allow its customers to place a share via their mobile phones.

WAP-based games form another WAP service that enables people to use mobile devices to gain access to a wider range of games while on the move (http://wirelessgames.com). A membership is required to access the full range of games available on the Web site. The Singapore-based portal of Oktopas (http://www.orktopas.com) provides news information that includes 4D and Toto result numbers, weather updates, and so on. The site also lists other WAP sites that are available in Singapore. The listings of the WAP sites are categorized in order to facilitate navigation and searching. The WAP site allows users to browse other WAP sites, such as Singapore eGuide and POEMS. In Singapore, WAP infrastructures and services are fully operational; however, the slow adoption of WAP and mobile devices is largely due to the limitations of mobile devices, as reported in an article in *Computer Times* (March 29, 2000).

# USE OF MOBILE DEVICES

A study was carried out in Singapore to investigate the extent to which mobile devices are used in e-commerce compared to fixed-network devices. Singapore was chosen due to the fact that it has one of the highest numbers of mobile devices users in the world. A number of factors were taken in consideration when conducting the survey. Some of these factors included user understanding and awareness of m-commerce, use of electronic devices, the features available in mobile devices, and the types of information and services accessed by different users. Some of the factors taken into consideration when comparing mobile devices with the fixed-network devices included sound effects, multimedia, user-friendliness of the user-interface, navigation, speed of access, user preferences, convenience of use, security, and size of the screen. It is also assumed that m-commerce users are computer literate, educated, and equipped with minimum knowledge of the WAP and mobile devices. The study involved 50 users in which 41 users responded to the survey. The other nine targeted were not available due to overseas engagements at the time of the study.

## Frequency and Usage of Features on Various Devices

**Mobile phones:** The study found that among the 41 respondents that participated in the survey, 68% use mobile phones daily, 15% use mobile phones at least once a week, while the other 17% does not use mobile phones at all (Table 1). None of the respondents fell under the category of mobile phone usage of at least once a month or more. Among the various devices that were used daily, mobile phone usage recorded the highest percentage (68%), followed by desktop usage (63%), followed by laptop usage, which made up 41%, and another 20% used PDAs. "Others" usage (included pagers and digital organizers) registered 22% of the total number of responses. It was clear

*Table 1: Frequency and Usage of All Devices*

| Frequency | Every day | At least once a week | At least once a month | Less often | Not at all |
|---|---|---|---|---|---|
| Mobile phone | 68% | 15% | 0% | 0% | 17% |
| PDA | 20% | 0% | 0% | 12% | 68% |
| Laptop | 41% | 3% | 3% | 7% | 46% |
| Desktop | 63% | 19% | 3% | 3% | 12% |
| Others | 22% | 0% | 0% | 3% | 76% |

that many users used more than one device, such as pagers, hand phones, laptops, and so on.

Among those who used mobile phones daily, 40% of the respondents used SMS and 7% accessed the Internet on a daily basis. We noticed that, apart from the basic functions of a hand phone (i.e., providing voice communication), users were using data communications features such as SMS and Internet. Many of the respondents may find sending messages (as compared to speaking over the phone) much more effective and cheaper, especially when the message is clear, and the conversation is short. Given that the cost of sending messages is low, and in many cases is free, for people on the move, the ability to perform simple online transactions (e.g., trading shares and getting weather forecasts) via the Internet with the help of mobile gadgets is attractive.

**PDAs:** The study found that PDAs are less used compared to hand phones (Table 2). Out of those who used PDAs daily, 55% cited calendaring as the most frequently used feature, while 18% used it as a calculator most of the time. The increasing popularity of PDAs is due to the switch from the manual recording of appointments on paper to using PDAs. PDAs provide multiple functionalities, such as calculations and recording of appointments. PDAs have been rather popular among the respondents simply because they can find a combination of features, such as calculators, address books, and games. Only 9% of the respondents used the Internet daily on their PDAs, which indicates that the Internet is still not the most popular feature used in PDAs. One of the possible reasons could be that such Internet features may not be available in the PDA models they are using, or the users are simply not aware of such feature. Other features specified by participants were scratchpads, spreadsheets, and notestakers on PDAs.

**Laptops:** Laptop usage, on the other hand, seems to be more popular. The study found that 41% of the respondents use laptops daily, 3% use them at least once a week, another 3% use them at least once a month, 7% less often,

*Table 2: Frequency and Usage of Features for PDA*

| Frequency | Every day | At least once a week | At least once a month | Less often | Not at all |
|-----------|-----------|----------------------|-----------------------|------------|------------|
| SMS | 9% | 18% | 9% | 0% | 64% |
| Infrared | 9% | 45% | 0% | 18% | 28% |
| Calendar | 55% | 18% | 9% | 9% | 9% |
| Games | 9% | 18% | 9% | 27% | 36% |
| Internet | 9% | 18% | 0% | 0% | 72% |
| Calculator | 18% | 45% | 9% | 9% | 18% |
| Others | 0% | 0% | 0% | 0% | 0% |

and 46% do not use the laptop at all (Table 3). The feature that recorded the highest daily usage rate by those surveyed (92%) was the Internet, and the next most-used feature was e-mail (80%). With the Internet and e-mail being the most desired features for laptop users, a similar trend as that of the desktop is shown. We expected such similarity because laptops and desktops provide basically the same functions, except for the fact that laptops are portable.

**Desktops:** Desktops are still the most popular platform for accessing the Internet. The study found that 63% of those surveyed use the desktop daily, 19% use the desktop at least once a week, 3% of the respondents use the desktop at least once a month, 3% use it less often, and only 12% do not use a desktop at all (Table 4). This shows that the desktop is the second most-used device as compared to the mobile phone. This is inevitable, as desktops are still the essential and conventional devices used in the office and at home. Out of the 63% who use the desktop daily, 75% access e-mail and 49% access the Internet daily. On a weekly basis, 27% and 9%, respectively, will access the Internet and e-mail. Besides e-mail and the Internet, the other features used and specified by respondents are word processors, spreadsheets, programming, and games.

*Table 3: Frequency and Usage of Features for Laptops*

| Frequency | Every day | At least once a week | At least once a month | Less often | Not at all |
|-----------|-----------|----------------------|-----------------------|------------|------------|
| E-mail | 80% | 4% | 0% | 4% | 8% |
| Infrared | 4% | 16% | 8% | 20% | 52% |
| Calendar | 28% | 12% | 4% | 20% | 36% |
| Games | 4% | 0% | 8% | 36% | 52% |
| Internet | 92% | 0% | 0% | 4% | 4% |
| Calculator | 4% | 20% | 4% | 52% | 20% |
| Others | 12% | 8% | 0% | 8% | 72% |

*Table 4: Frequency and Usage of Features for Desktop*

| Frequency | Every day | At least once a week | At least once a month | Less often | Not at all |
|---|---|---|---|---|---|
| E-mail | 75% | 9% | 6% | 6% | 3% |
| Calendar | 15% | 15% | 6% | 39% | 25% |
| Games | 12% | 27% | 15% | 30% | 16% |
| Internet | 49% | 27% | 12% | 6% | 6% |
| Calculator | 15% | 12% | 9% | 40% | 24% |
| Others | 3% | 0% | 0% | 3% | 94% |

# Use of Internet on Various Devices

Use of the Internet on the desktop had the highest number of respondents (64%), followed by the laptop (37%), and then the mobile phone (12%) (Table 5). In terms of user preference of devices in accessing the Internet, users still prefer to use the desktop 69%, followed by laptop 39%, then mobile phone 3%. This shows that the desktop is still the preferred device for daily use of the Internet.

Table 6 shows that the main purpose for accessing the Internet for most of the respondents is e-mail (90% of the respondents), followed by online news (85% of the respondents), and online shopping (46% of the respondents). This shows that e-mail became an essential communication tool. Internet news and online shopping are getting more popular. The less-preferred online services include accessing for online games (21% of respondents), restaurant/hotel reservations (18% of respondents), flight schedule checking (33% of respondents), as well as booking airlines or entertainment (28% of respondents).

Other reasons for accessing the Internet besides those mentioned above include participation in chat rooms or news groups (33% of respondents), conducting finance or banking transactions (41% of respondents), obtaining

*Table 5: Frequency of Usage of Internet on Various Devices*

| Frequency | Every day | At least once a week | At least once a month | Less often | Not all all |
|---|---|---|---|---|---|
| Mobile phone | 12% | 0% | 0% | 0% | 88% |
| PDA | 7% | 2% | 0% | 0% | 91% |
| Laptop | 37% | 5% | 0% | 5% | 53% |
| Desktop | 64% | 15% | 7% | 2% | 12% |
| Others | 0% | 0% | 0% | 0% | 0% |

*Table 6: Purposes for Accessing the Internet*

| Purpose of use | Number of respondents (%) |
|---|---|
| Participate in chat room or news group | 33 |
| Access e-mail | 90 |
| Conduct finance or banking transactions | 41 |
| Shop | 46 |
| Obtain stock quotes and purchase | 38 |
| Look for news | 85 |
| Look at advertisements | 38 |
| Look for jobs | 44 |
| Access music/movie clips | 41 |
| Book airline or concert/event/movie tickets | 28 |
| Make restaurant/hotel reservations | 18 |
| Obtain movie listings | 36 |
| Play games | 21 |
| Check flight schedules | 33 |
| Check weather | 31 |
| Others | 10 |

stock quotes and purchasing (38% of respondents), accessing online advertisements (38% of respondents), conducting online job hunts (44% of respondents), accessing music/movie clips (41% of respondents), obtaining movie listings (36% of respondents), checking the weather forecast (31% of respondents), and other online services (10% of respondents).

*Table 7: Priority of Usage of the Internet on Various Devices*

| Priority | 1 | 2 | 3 | 4 | 5 |
|---|---|---|---|---|---|
| Mobile phone | 3% | 12% | 6% | 0% | 0% |
| PDA | 0% | 0% | 12% | 3% | 0% |
| Laptop | 39% | 9% | 0% | 3% | 0% |
| Desktop | 69% | 27% | 3% | 3% | 0% |
| Others | 0% | 0% | 0% | 0% | 0% |

## WAP and Mobile Devices Usage Versus Fixed Network

For those who used the WAP on electronic gadgets, half of the respondents (50%) remain neutral with regard to the problem of navigation. However, the majority of the remaining half (33%) found no problem in navigating the Internet using mobile devices. The study also found that the majority of the respondents (50% of the WAP users) prefer to access the Internet via the desktop. Only 17% of the users preferred otherwise. Reasons pertaining to user interface, convenience of use, security, user-friendliness of user-interface, and speed of access, probably explain such a preference.

In terms of convenience, 33% of the respondents agreed that accessing the Internet via WAP and mobile devices is more convenient as compared to desktops and fixed networks. Another 33% remain neutral, while 34% disagreed. This indicates that the wireless network is more convenient for users to use to access information. On the other hand, 91% agreed that they do not have a problem navigating the Internet through fixed networks using a desktop. This indicates that respondents are more comfortable navigating the Internet using a desktop. Probably, WAP developers could find the points that desktop users are comfortable with and improve navigation on the mobile devices.

On security issues like how to secure the WAP compared to fixed networks, 50% of respondents were neutral, while 17% agreed, and 33% disagreed on whether security is an issue in accessing the Internet. The majority of respondents, 47%, strongly agreed, and 39% agreed that security is an issue on the Internet. This pointed to the need for WAP developers and implementers to improve on the security so as to attract WAP users or even non-WAP users to use wireless technology for accessing the Internet. However, the majority of respondents (61% of the respondents) remains neutral regarding whether using a desktop network to access the Internet is more secure. Of them, 6% strongly agree, 25% agree, and 8% disagree. This indicates that users are still more inclined toward the thinking that using a desktop network to access the Internet is more secure.

With regard to user-friendliness of WAP and mobile devices, 66% of the users agreed that user interfaces are not friendly and are hard to use on mobile devices. Only 17% agreed that the user interfaces are friendly, while the remaining 17% were neutral. This suggests that a lot of work needs to be done to improve user interfaces on mobile devices. On the comparison of the speed acceptability for Internet access through WAP and fixed networks, 33% of the respondents agreed, while 33% disagreed. This shows that users at this stage are not much worried about speed. This may be due to the fact that WAP and mobile devices still have limited functionality, and users do not see the need for faster services. With regard to respondents' opinions on the need for further software development to improve WAP and mobile devices services, 66% agreed. This shows that a lot of work still needs to be done to meet the users' needs.

The designs of the different electronic gadgets are taken into consideration as well. Most of the respondents, 56%, agreed that the size of the screen on mobile devices is still not acceptable. The larger percentage that disagreed found that the screen size on mobile devices imposed many restrictions on the functionality and the user-interface. There are, however, 33% of respondents who think that the screen size is an issue, and they can live with it. The majority of those surveyed 59% did not agree that the sound effects on mobile devices are acceptable. They all indicated that further improvement is needed. This is the same with multimedia capabilities, in which 71% agreed that mobile devices are yet to catch up with those available on desktops and fixed networks.

# CONCLUSION

Mobile devices are increasingly getting popular, especially with the younger generation who wants fast and convenient services. These services include messaging, gaming, making financial transactions, shopping, and downloading music and movies. But while mobile devices have great potential in e-commerce, limitations, such as screen size, small keyboards, little memory, and lower processing powers, hindered the use of such devices for serious e-commerce transactions. From the usability study carried out in Singapore, we found that more than half of the e-commerce users (54%) are aware of the WAP and wireless technology and are also familiar with mobile devices. The study also shows that most users utilize the Internet for accessing e-mail, online news, and sometimes online shopping. Not all e-commerce users use WAP for e-commerce transactions despite the fact that most of their mobile devices are

WAP enabled. The study showed that most users prefer to use desktop for Internet access, citing that size of the screen, speed, stability, and security are some of the reasons. E-commerce users in Singapore view security to be an issue in both fixed-network and wireless technology. They have not completely adopted and embraced the wireless technology, as the immediate benefits may not be that apparent yet. There is still a lot of development work to be done to improve the limitations of mobile devices and network bandwidth.

Despite all of these limitations, wireless technology and mobile devices are still attractive and have great potential in m-commerce. Fast and easy access to information is becoming important for people on the move. Convenience of access is also one of the many attributes of mobile devices that consumers are looking for.

# REFERENCES

Arthur Andersen. (2002). Are you ready? The use of mobile data applications in Europe (pp. 1-7). Retrieved October 10, 2002, from the World Wide Web: www.iora.com/downloads/Andersen.pdf.

Chin, J. & Chua, H. H. (2000). Bullish about hi-tech trading. *Computer Times*, (April 5).

Cloberg, T. P. (1995). *The Price Waterhouse EDI Handbook*. New York: John Wiley & Sons.

Compton, M. (2002). M is for mobile. [Online]. Retrieved October 12, 2002 from the World Wide Web: http://dir.salon.com/tech/view/2000/01/31/wap/index.html.

Emanuel, S. (2000). The crawl of m-commerce. *Computerworld,* 6(47). Retrieved October 12, 2002 from the World Wide Web: http://computerworld.com.sg/pcwsg.nsf/unidlookup/CA99558961F50FBC48256B4F002AE396?OpenDocument.

Furuskar, A., Naslund, J., & Olofssun, H. (1999). Edge-enhanced data rates for GSM and TDMA/136 Evolution. *Ericsson Review*, 1, 28-37.

Garcia, M. R. (2000, May 22-28). Companies cut the cord. *Asia Computer Weekly*, 21(17). Retrieved October 12, 2002, from the World Wide Web: http://www.informationweek.com/785/wireless.htm.

Loken, S. B. (2001). Trivial Pursuit goes mobile. Retrieved March 31, 2001, from the World Wide Web: http://www.wap.com.

Oo, G. L. (2000). Ordering in fast food the WAP way. *The Straits Times*, (May 3).

Reed, M. (2000). Why the future will be wireless. [Online]. Retrieved October 10, 2002, from the World Wide Web: http://www.mobileresearchers.com/contents.asp?art=39.

Rudkin, C. (2001). Fun in your mail. Retrieved March 31, 2001 from the World Wide Web: http://www.wap.com.

Senthilnathan, R. (2000). Singapore is 10th highest in Net usage. *The Straits Times,* (March 29).

Tee, E. (2000). Regional e-commerce yet to take off in a big way. *The Straits Times*, (March 3).

Unified messaging services for SMEs with WAP (2000). *Computer Times*, (March 29).

U.S. mobile devices to 2006, A land of opportunities (2002). Datamonitor, 2001. Retrieved October 10, 2002, from the World Wide Web: http://www.mobilecommerceworld.com/EDriveFiles/DataMonitor/11_RPDF.pdf.

Young, D. (1995). Considerations in the design of electronic commerce solutions, part 1. *Electronic Commerce World*, 8. Retrieved October 10, 2002, from the World Wide Web: http://wint.decsy.ru/internet/digital/v0000257.htm.

# About the Editor

**Nan Si Shi** received his Ph.D. in Information Systems Management at the University of South Australia, and his Master's in Computer Networks at Nanyang Technological University Singapore. Dr. Shi has more than 20 years of experience in the Information Systems field, including industry practice, academic research, and teaching. He participated extensively in research projects on information technologies. He has authored and edited a few books, *Essential Technologies for E-Commerce*, *Information Technology of E-Commerce*, and *Architectural Issues of Web-Enabled Electronic Business*, and has published a number of research papers and chapters. He is currently responsible for the areas of corporate IT strategy planning, e-business, m-commerce, IT security policy, information management, innovation management, etc. He is also Adjunct Research Associate of the Division of Business and Enterprise at the University of South Australia, and a member of the International Board of Editors for the *Journal of Information Technology Education*. He is a member of the advisory board for University of South Australia Bachelor of Engineering in Singapore. His recent teaching experience includes Information Systems Management for MBA.

# About the Authors

**Suliman Al-Hawamdeh** is Founder and Program Director of the Master of Science in Knowledge Management at Nanyang Technological University (Singapore). He is also Founder and President of the Information and Knowledge Management Society (iKMS), a professional organization dedicated to the advancement of the information and knowledge management profession. Dr. Hawamdeh is the author of two books on knowledge management, *Information and Knowledge Society* (McGraw-Hill) and *Knowledge Management: Cultivating the Knowledge Professionals* (Chandos Publishing, Oxford). He is the Editor-in-chief of the *Journal of Information and Knowledge Management*.

**C.R. Chatwin**, Professor, holds the Chair of Industrial Informatics and Manufacturing Systems (IIMS) at the University of Sussex, UK; where, *inter alia*, he is Director of the South East Advanced Technology Hub (SEATH), the IIMS Research Centre, and the Laser and Photonic Systems Research Group. Before moving to Sussex, Professor Chatwin spent 15 years at the University of Glasgow, Engineering Faculty (Scotland, UK), where, as a Reader, he was head of the Laser and Optical Systems Engineering Centre and Industrial Informatics Research Group. He published two research-level books—one on numerical methods, and the other on hybrid optical/digital computing—and more than 200 international papers.

**Constantinos Coursaris** is a Ph.D. student in the Information Systems area at the DeGroote School of Business at McMaster University (Ontario, Canada). His Ph.D. research focuses on e-business and mobile commerce. He holds an MBA in e-business from McMaster University, and a B.Eng. in Aerospace from Carleton University (Ontario, Canada). He has held management positions in information technology, retail, and hospitality. He has several publications in the area of mobile commerce, including articles in the *Canadian Journal of Administrative Studies* and the *Quarterly Journal of Electronic Commerce*. His research interests include e-business, m-commerce, wireless privacy, wireless security, usability of wireless devices, wireless applications, m-commerce pricing, and location-based services.

**Nikhilesh Dholakia** is a Professor in the marketing, e-commerce, and management information systems areas at the University of Rhode Island (URI), USA, and a faculty associate at URI's Research Institute for Telecommunications & Information Marketing. He has published extensively in the fields of marketing, e-commerce, and consumer culture, and taught in several academic and executive programs in Asia, Europe, and Latin America. Dr. Dholakia won the Charles Slater award of the *Journal of Macromarketing*. He also chaired doctoral dissertations that won the Marketing Science Institute's Alden G. Clayton award and the Association for Consumer Research/Sheth Foundation award, and he supervised award-winning student essays at the Pacific Telecommunications Council. His recent books include *Worldwide E-Commerce and Online Marketing: Watching the Evolution* (Quorum, 2002) and *M-Commerce in North America, Europe, and Asia-Pacific: Cases and Readings* (Idea Group Publishing, 2003, forthcoming). Dr. Dholakia holds a B.Tech. in Chemical Engineering from the Indian Institute of Technology, an MBA from the Indian Institute of Management, and a Ph.D. from the Kellogg School of Management at Northwestern University (Illinois, USA).

**Ruby Roy Dholakia** is Director of the Research Institute for Telecommunications and Information Marketing (RITIM) in the College of Business Administration at the University of Rhode Island (URI) (USA) and Founder of the COTIM series of conferences. She is also a Professor of Marketing at URI. She holds a B.S. in Marketing and an MBA from the University of California at Berkeley and a Ph.D. in marketing from Northwestern University (Illinois, USA). Engaged extensively in research projects on telecommunications and information technologies for the home and the workplace, she authored numerous books, such as *Marketing Strategies for Information Technolo-*

*gies* (JAI Press, 1994), *New Infotainment Technologies in the Home: Demand-Side Perspectives* (Lawrence Erlbaum Associates, 1996), and *Worldwide E-Commerce and Online Marketing: Watching the Evolution* (Quorum, 2002). Her research on information technology consumers and markets, including e-commerce and m-commerce topics, appeared in major business and IT journals.

**Sheng-Uei Guan** received his M.Sc. and Ph.D. from the University of North Carolina at Chapel Hill (USA). He is currently with the Electrical and Computer Engineering Department at National University of Singapore. Professor Guan also worked in a prestigious R&D organization for several years, serving as a design engineer, project leader, and manager. He also served as a member on the R.O.C. Information & Communication National Standard Draft Committee. After leaving the industry, he joined Yuan-Ze University in Taiwan for three and a half years. He served as Deputy Director for the Computing Center, and also as Chairman for the Department of Information and Communication Technology. Later he joined La Trobe University (Melbourne, Australia) with the Department of Computer Science and Computer Engineering, where he helped to create a new multimedia systems stream.

**Khaled Hassanein** is Associate Professor of Information Systems at the DeGroote School of Business at McMaster University (Ontario, Canada). His current research interests are in the areas of e-business and m-commerce, including online trust, e-finance, e-health, automated website analysis and personalization using pattern recognition techniques, wireless device usability, as well as online privacy. Dr. Hassanein has published in the *Quarterly Journal of Electronic Commerce, Canadian Journal of Administrative Sciences, Journal of Information Technology Cases and Applications*, and the *Neural Networks Journal*, among others. He has presented at many international conferences related to e-business and pattern recognition. Dr. Hassanein is a Senior Member of the IEEE and a Co-director of the annual McMaster eCase Competition.

**Milena Head** is the Director of the McMaster eBusiness Research Centre (MeRC) (Canada), Faculty Director for the World Congress on the Management of Electronic Business, and a Co-director of the annual McMaster eCase Competition. Specializing in e-business and HCI, she has published in *International Journal of Human–Computer Studies, Interacting with Computers, Group Decision and Negotiation, Internet Research, Human Systems*

*Management, Quarterly Journal of Electronic Commerce, Journal of Business Strategies, Canadian Journal of Administrative Sciences*, among others, and has presented at numerous international conferences. Her research interests include trust and privacy in electronic commerce, interface design, mobile commerce, Web navigation, Web-based agents, supply chain collaboration, online negotiation, e-retailing, and information retrieval.

**Nir Kshetri** is Assistant Professor in the College of Business at the University of North Carolina at Greensboro, USA (from August 2003). Previously, he was on the faculty of Management School, Kathmandu University (Nepal). His research focuses on international e-business and technology diffusion. His works were published in *Small Business Economics, Electronic Markets, Pacific Telecommunications Review*, and as chapters in several books and conference proceedings. He was the winner of the 2001 Association for Consumer Research/Sheth Foundation Dissertation Award. He also won first prize in the Pacific Telecommunications Council Essay Competition in 2001 and second prize in the same competition in 2000. He obtained his Ph.D. from the University of Rhode Island.

**Bonnie Lam** graduated cum laude from the University of Cape Town (South Africa) in 2002, achieving her Bachelor of Science Honors degree in Computer Science. During her studies, she focused on the networks and databases areas. She is interested in communication technology and distributed applications. Bonnie Lam is currently employed by UUNET SA (Pty) Ltd. as one of the developers in the company.

**Mark Lehrer** is Assistant Professor of Management at the University of Rhode Island, USA. After obtaining his Ph.D. at INSEAD, he worked as a Research Fellow at the Social Science Center Berlin (WZB; Germany) in 1997 and 1998. His research interests revolve around comparative national management styles, European civil aviation, globally distributed knowledge management, and German high-tech. Recent publications appeared in *Journal of World Business, Organization Studies, Industrial and Corporate Change* and *California Management Review*.

**P.W. Lei** is a Ph.D. student in the School of Engineering and Information Technology at the University of Sussex, UK. Her research interests include multiagent systems, auction market model, evolutionary computing, electronic commerce, and management of information technology.

**Ken MacGregor** is the Professor of Computer Science and Head of the Computer Science Department at the University of Cape Town in South Africa. His research focus is predominately distributed computing systems and distributed application development, particularly wireless distributed systems. Ken MacGregor has been a Full Professor at the University of Cape Town since 1974, during which time he has been involved with many commercial organizations, the latest being Vodacom, the Vodaphone subsidiary in South Africa, where he has acted as a consultant.

**Chon Seng Ngoo** received his B.S. from National University of Singapore in 2002. His research interests include software agents and electronic commerce.

**John H. Nugent**, D.B.A., C.P.A., serves as an Assistant Professor in the Graduate School of Management (GSM) at the University of Dallas (Texas, USA), where he teaches strategy, wireless, and other telecommunications courses. He also established GSM's Information Assurance concentration and serves as the Director of the University's Center of Information Assurance. He concurrently serves as CEO of the Hilliard Consulting Group, Inc., a leading strategy consulting firm in the telecommunications and IT industry segments. Previously, he served as President and a Board of Director member of a number of AT&T subsidiaries. John was awarded the *Defense Electronics* "10 Rising Stars" award in July 1989 as well as the Diplome de Citoyen D'Honneur, Republic of France, in June 1988.

**Tommi Pelkonen** is a Doctorate Candidate at the Helsinki School of Economics (Finland). In his professional career, he worked as Management Consultant specializing in mobile telecommunications, internationalization, and business strategy formulation at Satama Interactive (www.satama.com), a European digital services firm. He worked on multiple mobility-related business projects. Prior to Satama, Tommi Pelkonen worked as project manager and researcher in LTT-Research Ltd. (www.ltt-tutkimus.fi), analyzing the developments in the Finnish interactive service provision markets. This topic also forms the theme of his doctoral dissertation. M.Sc. (Econ.) Pelkonen authored several publications of the Finnish digital media landscape. His latest report (2002) was on analyzing the earnings logic of the Finnish digital television industry. In addition, Mr. Pelkonen worked as IT project supervisor and lecturer in the Information Technology Program (ITP) at the Helsinki School of Economics.

**Mahesh S. Raisinghani** is Program Director of eBusiness and a faculty member at the Graduate School of Management, University of Dallas (Texas, USA), where he teaches MBA courses in Information Systems and eBusiness. Dr. Raisinghani was the recipient of the 1999 UD Presidential Award; 2001 King/Haggar Award for excellence in teaching, research, and service; and the 2002 research award and a finalist at the 2002 Asian Chamber of Commerce awards. He serves as an Associate Editor and on the editorial review board of leading information systems/e-commerce journals and on the board of directors of Sequoia, Inc. Dr. Raisinghani is included in the millennium edition of *Who's Who in the World, Who's Who Among America's Teachers*, and *Who's Who in Information Technology.*

**Mats Samuelsson** worked within the telecommunications industry since the early 1980s, in areas spanning traditional telephony as well as broadband systems and services. With experience in product management, business development, and marketing for US and international markets, he worked with service providers all over the world, introducing new service technology for telephone broadband and wireless networks. Service technologies include Intelligent Networks, messaging, IP, and other enhanced services. He is currently working on setting up a venture focused on allowing wireless operators to offer integrated business services to their business customers. Mats Samuelsson holds graduate degrees in business administration, manufacturing engineering, and electrical engineering from the Wharton School, Boston University (Massachusetts, USA), and the Royal Institute of Technology in Stockholm, Sweden, respectively.

**S.H. Tóng** received a B.BA. degree in Business Information Systems and an MBA degree in Finance and Banking from the University of Macao, China, in 1994 and 1998, respectively. She was Lecturer of Quantitative Methods of this university. At present, she is a Ph.D. student in management science at Instituto Superior de Ciências do Trabalho e da Empresa, Lisbon, Portugal. Her research areas are information systems, logistics, electronic commerce, and banking.

**Nico de Wet** is a Master of Science student in the Data Network Architecture Lab at the Computer Science department of the University of Cape Town, South Africa. His research is focused on software performance engineering using the UML 2.0 standard. Nico de Wet graduated cum laude from the University of Cape Town in 2002, achieving his Bachelor of Science Honors degree in Computer Science. He has an active interest in communication technology, wireless middleware, and enterprise application integration.

**Nadim Yazdani** completed a Bachelor of Science (Information Systems) degree at Rhodes University, Grahamstown (South Africa), in 2001 before proceeding to do his Honours in Computer Science at the University of Cape Town (South Africa) the following year. Nadim Yazdani currently works as a Software Analyst/Developer for a Cape Town-based software company specializing in the delivery of novel Web applications to a variety of local and international clientele.

**R.C.D. Young** obtained his undergraduate and Ph.D. degrees from Glasgow University (Scotland, UK). Until 1993, he was employed within the Laser and Optical Systems Engineering Research Centre at Glasgow, during which time he gained wide experience in optical systems engineering and image/signal-processing techniques. He participated in two European-funded electro-optical projects involving pan-European collaboration between leading European Universities and Industry. The second of the projects was proposed and led by Glasgow University. In April 1995, he was appointed a Lecturer in the School of Engineering at the University of Sussex (UK), a Senior Lecturer in October 1998, and a Reader in October 1999. There, he is continuing research into various aspects of optical pattern recognition, digital image processing, and electro-optics system design, and he is applying this to a wide range of problems of industrial relevance. More than 70 of his publications appear in peer-reviewed academic journals and at international conferences, many of them invited as papers to special issues. And, he has been invited as a keynote speaker to several conference sessions. He chairs sessions in the conference on Optical Pattern Recognition held each year by SPIE in Orlando, Florida, USA. He is a member of the Society of Photo-Optical Instrumentation Engineers (SPIE), the Optical Society of America, and the IEEE.

**Fangming Zhu** received his B.S. and M.S. degrees from Shanghai Jiaotong University, China, in 1994 and 1997, respectively. After graduation, he joined Shanghai Ricoh Facsimile Co. Ltd. as a Research Engineer. He is now a Ph.D. candidate in the Department of Electrical and Computer Engineering at National University of Singapore. His current research interests include intelligent agents, evolutionary computation, and agent-based electronic commerce.

# Index

WMLScript  248
World Wide Web (WWW) Model  85

## X

XML middleware  221

## Y

Y2K  188

# *NEW* from Idea Group Publishing

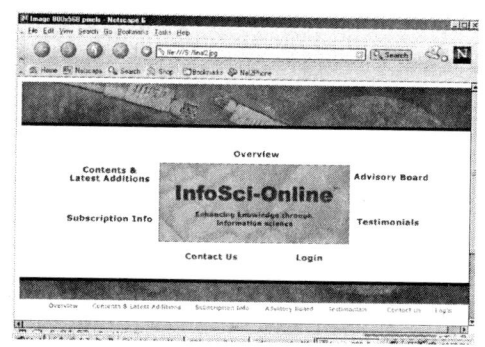

# Journal of Electronic Commerce in Organizations (JECO)

## The International Journal of Electronic Commerce in Modern Organizations

| | |
|---|---|
| **ISSN:** | 1539-2937 |
| **eISSN:** | 1539-2929 |
| **Subscription:** | Annual fee per volume (4 issues): Individual US $85 Institutional US $205 |
| **Editor:** | Mehdi Khosrow-Pour, D.B.A. Information Resources Management Association, USA |

## Mission

The *Journal of Electronic Commerce in Organizations* is designed to provide comprehensive coverage and understanding of the social, cultural, organizational, and cognitive impacts of e-commerce technologies and advances on organizations around the world. These impacts can be viewed from the impacts of electronic commerce on consumer behavior, as well as the impact of e-commerce on organizational behavior, development, and management in organizations. The secondary objective of this publication is to expand the overall body of knowledge regarding the human aspects of electronic commerce technologies and utilization in modern organizations, assisting researchers and practitioners to devise more effective systems for managing the human side of e-commerce.

## Coverage

This publication includes topics related to electronic commerce as it relates to: Strategic Management, Management and Leadership, Organizational Behavior, Organizational Developement, Organizational Learning, Technologies and the Workplace, Employee Ethical Issues, Stress and Strain Impacts, Human Resources Management, Cultural Issues, Customer Behavior, Customer Relationships, National Work Force, Political Issues, and all other related issues that impact the overall utilization and management of electronic commerce technologies in modern organizations.

**For subscription information, contact:**

Idea Group Publishing
701 E Chocolate Ave., Ste 200
Hershey PA 17033-1240, USA
cust@idea-group.com
URL: www.idea-group.com

**For paper submission information:**

Dr. Mehdi Khosrow-Pour
Information Resources Management Association
jeco@idea-group.com